Shallow-Level Plutonic Complexes in the Eastern Sierra Nevada, California, and Their Tectonic Implications

RICHARD A. SCHWEICKERT
Department of Geology and
Lamont-Doherty Geological Observatory of Columbia University
New York, New York 10027

THE
GEOLOGICAL SOCIETY
OF AMERICA

SPECIAL PAPER 176

The Special Paper series was originally made possible through the bequest of Richard Alexander Fullerton Penrose, Jr.

Library of Congress Catalog Card Number 76-11570
I.S.B.N. 0-8137-2176-8

Published by
The Geological Society of America, Inc.
3300 Penrose Place
Boulder, Colorado 80301

Printed in The United States of America

Contents

ILLUSTRATIONS

Acknowledgments

I am deeply indebted to W. R. Dickinson, under whose guidance, encouragement, and support this study was undertaken. Discussions with him, R. R. Compton, D. S. Cowan, W. C. Luth, B. M. Page, and the late C. O. Hutton played an important part in developing the ideas presented. The manuscript benefited greatly from suggestions by W. R. Dickinson and D. S. Cowan (an earlier version) and by W. Alvarez, I.W.D. Dalziel, R. W. Kistler, and O. T. Tobisch.

Field and laboratory work was supported by National Science Foundation Graduate Fellowships (1968-71), Geological Society of America Grants 1309-69 and 1393-70, and the Shell Fund at Stanford University. Chemical analyses were provided by Alberto Obregón of the Instituto de Geologia, Universidad Nacional Autónoma de México; R. L. Armstrong kindly provided K-Ar analyses. Ruperto Laniz prepared more than 200 thin sections. Texaco Inc., and Lamont-Doherty Geological Observatory provided much assistance in drafting figures.

This paper is part of a Ph.D. dissertation completed at Stanford University in 1972.

Abstract

Several related sequences of plutonic and metavolcanic rocks of probable Early Jurassic age occur between Lake Tahoe and Mono Lake in the eastern Sierra Nevada and Sweetwater Range, California. The plutons solidified at very shallow levels; some broke through to the surface to erupt compositionally similar pyroclastic rocks and lava flows. In many respects, the more felsic plutons resemble Tertiary epizonal plutons in the Northern Cascade Mountains.

The pre-Cretaceous volcano-plutonic complexes are part of a terrane of shallow-level mafic and felsic intrusions and remnants of metamorphosed sedimentary and volcanic rocks, all of probable Late Triassic to Middle Jurassic age, that extends the length of the Sierra Nevada in eastern California and Nevada. Felsic plutons of this general age typically have seriate textures and are moderately metamorphosed. They form a low-quartz and high-K_2O group that is modally and chemically distinct from Upper Jurassic and Cretaceous plutons that occur locally among them and that also form the main expanse of the Sierra Nevada batholith to the west. The extent and arrangement of this terrane suggest that it is part of an Andean-type magmatic arc that developed between Late Triassic and Middle Jurassic time in eastern California and western Nevada, generally east of the site of the second, much larger, magmatic arc that developed mainly in Cretaceous time. Triassic and Jurassic subduction linked to the evolution of this older arc probably occurred in the foothills of the western Sierra Nevada. Later subduction of the Franciscan Complex in the California Coast Ranges was linked to Sierra Nevada plutonic activity that occurred after Middle Jurassic time.

Introduction

Plutonic rocks of the Sierra Nevada and adjacent areas in eastern California and western Nevada are considered by many to be the subvolcanic part of a major west-facing magmatic arc of Andean type that developed on older Paleozoic basement rocks at the west edge of the North American continent during most of the Mesozoic Era (Hamilton, 1969b; Dickinson, 1970). From west to east, granitic plutons in the Sierra Nevada batholith range in age from Late Jurassic to Late Cretaceous (Evernden and Kistler, 1970). In this interpretation, these plutons are the arc rocks related to coeval sedimentary and volcanic rocks of the Franciscan Complex and Great Valley sequence, the so-called trench-subduction zone and arc-trench gap sequences, respectively.

However, much older plutons, with apparent ages ranging from Late Triassic to Middle Jurassic, lie farther east in the eastern Sierra Nevada, the White and Inyo Mountains, and the Argus Range (Evernden and Kistler, 1970) and clearly disrupt the simple pattern in which progressively younger intrusions are encountered from west to east. These Upper Triassic to Middle Jurassic plutons (called "older plutons" in this paper) cannot be easily accounted for in the above model, because they have no counterparts in sedimentary rocks or structures in the California Coast Ranges; furthermore, they are anomalous for reasons other than their old ages with respect to their geographic position. For example, the older plutons are spatially associated with remnants of probably coeval volcanic and sedimentary rocks. These volcanic and sedimentary rocks occur primarily in roof pendants and as wall rocks of younger plutons. The younger Upper Jurassic to Upper Cretaceous plutons, on the other hand, have no known preserved eruptive or volcanic counterparts in the Sierra Nevada region. This fact suggests that the older plutonic province paradoxically lies at a shallower level of exposure than adjacent plutons that are as much as 100 m.y. younger and whose presumed former volcanic cover has been entirely stripped off and shed into the flanking Great Valley sequence to the west during Late Jurassic to Late Cretaceous time (Ojakangas, 1968; Dickinson and Rich, 1972).

To gain more insight into the geologic relationships of the older plutonic province, several of the older plutons and certain metavolcanic units near

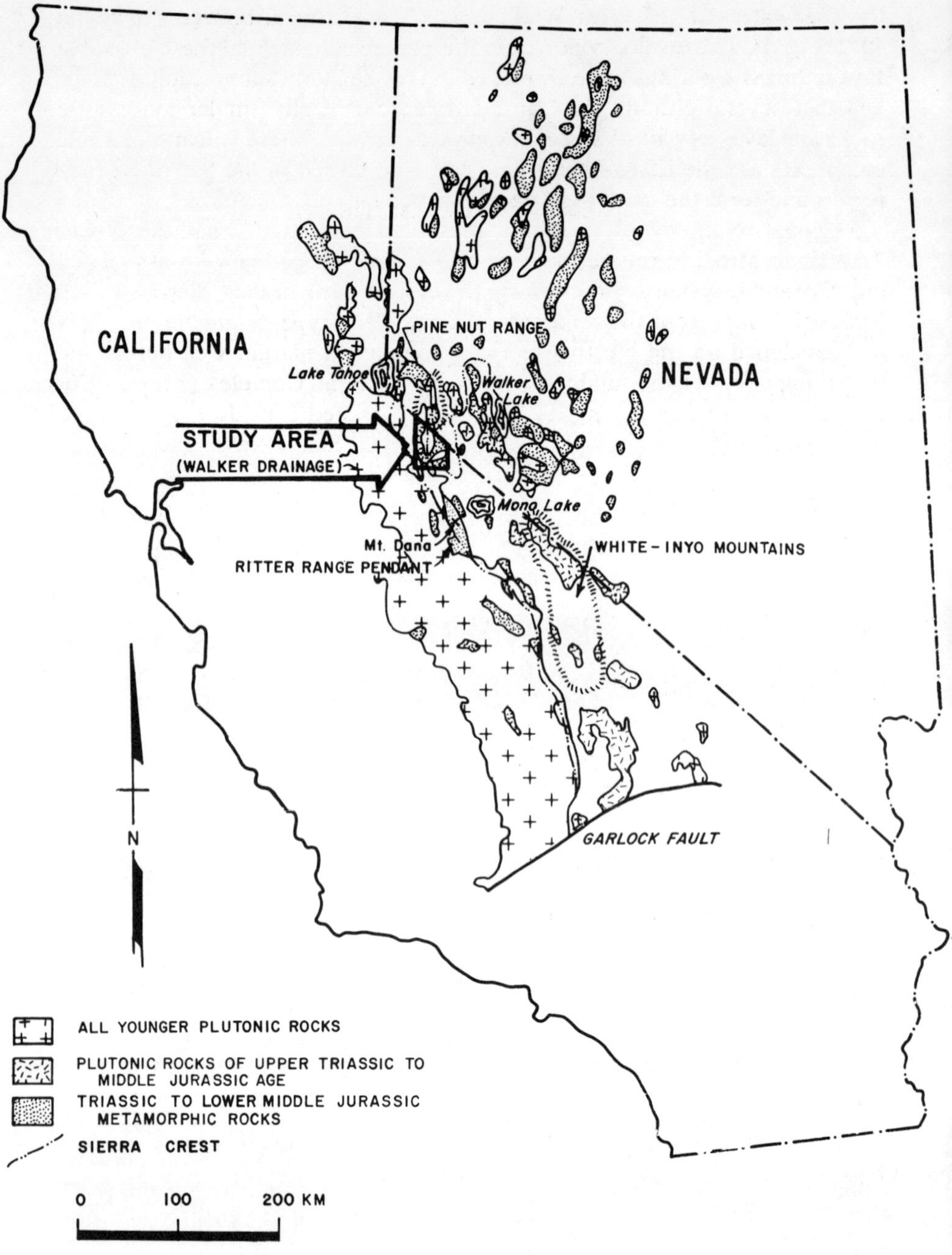

Figure 1. Index map of California and Nevada, showing the Sierra Nevada batholith, the Sierra Nevada crest, and the outline of the area studied.

the headwaters of the West Walker River were studied between 1968 and 1972 (Fig. 1). The results, reported in this paper, indicate that these probable Lower Jurassic plutons were emplaced in very shallow, subvolcanic settings and that several plutons may have erupted chemically similar pyroclastic rocks and lavas as part of an accreting volcanic pile. These volcano-plutonic complexes are the first of their kind to be described in the Sierra Nevada region and form the main topic of this paper.

Regional relationships and data presented here argue that the Upper Triassic to Middle Jurassic plutonic rocks, which together with associated eruptive and sedimentary rocks form the eastern flank of the "Sierra Nevada batholith," represent the remnants of an Andean-type magmatic arc. This arc developed on the North American continental margin well before the beginning of either the subduction of the Franciscan Complex (Ernst, 1970; Suppe, 1969, 1973; Cowan, 1974) or the associated Late Jurassic to Late Cretaceous magmatic activity that produced the "Sierra Nevada batholith." In other words, there is evidence that the Sierra Nevada region records the superposition of two distinct Andean-type arcs during Mesozoic time, the older having been buried and contact metamorphosed by the younger.

REGIONAL SETTING

Bateman and Dodge (1970) delineated eight sequences of intrusive rocks in the central Sierra Nevada between lat 36°45′ and 38°00′N. They thought that each sequence was emplaced mainly during one of five major intrusive pulses recognized by Evernden and Kistler (1970) in a continuum of ages from about 210 to 80 m.y. B.P., although the existence of these pulses has recently been challenged (Lanphere and Reed, 1973). Together, these intrusive sequences apparently form elongate belts that are subparallel to the axis of the batholith (Bateman and Dodge, 1970; Evernden and Kistler, 1970). Rocks of the oldest sequences, with ages ranging from Late Triassic to Middle Jurassic, occur mainly east of the Sierra Nevada crest in the eastern Sierra Nevada and in the White and Inyo Mountains (Fig. 1). Flanking these rocks on the west are the Upper Cretaceous plutons that occupy the axial part of the batholith. The oldest plutons that I have studied in the drainage area of the West Walker River (Fig. 2) are intruded by several of the Upper Cretaceous plutons, and although they possess "young" (76 to 93 m.y.) apparent K-Ar ages (discussed below), several lines of evidence suggest these plutons are probably of Early Jurassic age. As several investigators have pointed out, K-Ar ages of older plutonic rocks in the eastern Sierra Nevada are difficult to interpret because most of the older plutons have experienced complex subsolidus thermal histories during intrusion of younger plutons, and discordant dates on mineral pairs

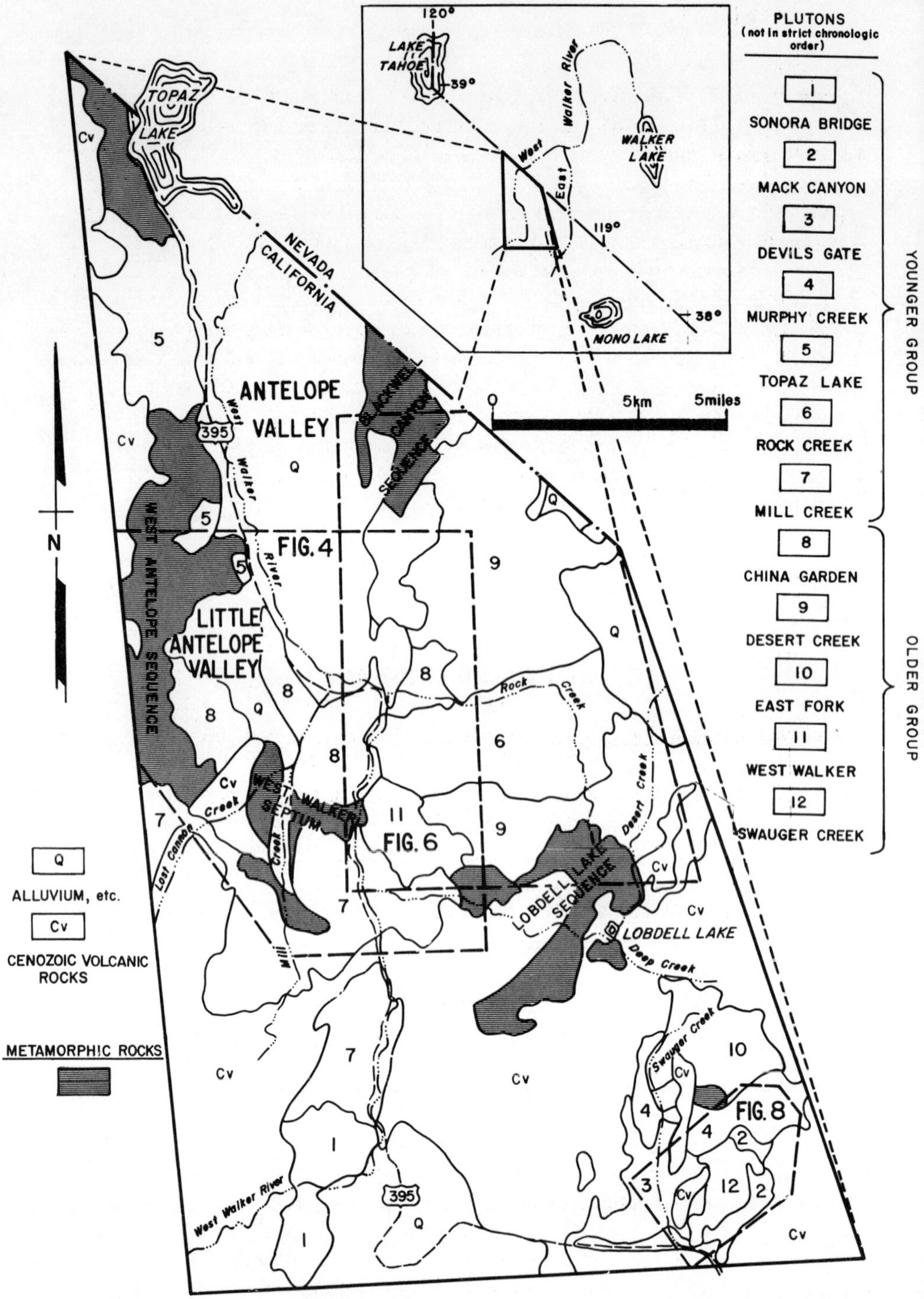

Figure 2. Key map for plutons and metamorphic sequences in the area studied. Also shown are locations of Figures 4, 6, and 8.

TABLE 1. PETROGRAPHIC AND FIELD DATA FOR PLUTONS OF THE WALKER DRAINAGE (SEE FIG. 2)

	Pluton	Age	Chief rock type	Average modes*				N†
				Q	K	P	M	
1.	Sonora Bridge	Younger than 7	Biotite granite; medium-grained, allotriomorphic granular	32.7	42.5	20.8	3.8	2
2.	Mack Canyon	Younger than 4 and 12; 79 m.y.	Biotite granite; medium-grained, allotriomorphic granular	30.8	41.9	24.4	2.3	3
3.	Devils Gate	Younger than 4 and 12; 83 m.y.	Biotite granite; coarse-grained, allotriomorphic granular	30.5	42.6	21.3	4.7	5
4.	Murphy Creek	Younger than 10 and 12	Biotite granite; medium- to coarse-grained, hypidiomorphic granular	30.0	40.0	20.0	10.0	3
5.	Topaz Lake	82 m.y. (Evernden and Kistler, 1970)	Hornblende-biotite quartz monzonite; coarse-grained, porphyritic, hypidiomorphic granular	24.4	27.0	40.7	7.6	6
6.	Rock Creek	Younger than 8, 9, and 11	Hornblende-biotite quartz monzonite; coarse-grained, porphyritic, hypidiomorphic granular	25.5	25.0	41.7	6.3	8
7.	Mill Creek	Younger than 11, older than 1	Biotite quartz monzonite; coarse-grained, porphyritic, hypidiomorphic granular	22.3	27.7	41.8	8.1	9
8.	China Garden	Younger than 9 and 11, older than 6	Biotite-hornblende quartz monzonite; medium- to coarse-grained, seriate	21.6	28.2	41.4	7.2	7
9.	Desert Creek	Younger than 11, older than 6 and 8	Augite-hornblende quartz monzonite; coarse-grained, seriate	19.2	32.6	35.6	10.7	7
10.	East Fork	Older than 4	Biotite-hornblende quartz monzonite; medium- to coarse-grained, seriate	22.4	28.8	34.0	14.4	7
11.	West Walker	Older than 6, 7, 8, and 9	Diorite and quartz diorite; medium- to coarse-grained, subophitic	Color index 28 to 40				
12.	Swauger Creek	Older than 2, 3, 4, and 10	Gabbro and olivine gabbro; medium- to coarse-grained, ophitic	Color index 40 to 88				

*1 through 7 and 10: points counted on stained slabs; average 600 points counted. 8, 9, 11, and 12: points counted in thin sections; average 600 points counted. Q, quartz; K, K-feldspar; P, plagioclase; M, mafic minerals.

†N is number of modally analyzed samples.

are the rule rather than the exception (Evernden and Kistler, 1970; Crowder and others, 1973; Lanphere and Reed, 1973).

Sporadic exposures of deformed metavolcanic and metasedimentary rocks of Late Triassic to Middle Jurassic age occur in a chain of roof pendants that is parallel to the known extent of Upper Triassic to Middle Jurassic plutonic rocks and extends over 500 km from the northernmost Sierra Nevada to the Mojave Desert (Figs. 1 and 11). Throughout their extent, the metavolcanic and metasedimentary rocks generally lie a few kilometres west of the belt of older plutons, but in the Walker River drainage area, as shown in this paper, the older plutons and coeval volcanic rocks are intimately associated.

Metavolcanic and metasedimentary rocks that crop out in the vicinity of the West Walker River are lithologically and structurally similar to fossiliferous Upper Triassic to upper Lower Jurassic strata in the southern Pine Nut Range (Noble, 1962; D. C. Noble, 1969, 1970, oral communs.), which is 25 km to the north, and to metasedimentary and metavolcanic rocks exposed near the Virginia Lakes and on Mount Dana in the Ritter Range pendant (Kistler, 1966a, 1966b), which is 40 to 60 km to the south (Figs. 1 and 11). Early to Middle Jurassic fossils were reported from near Lobdell Lake in the West Walker River area (in the Lobdell Lake sequence of this paper; Fig. 2) by Halsey (1953), but no fossils were found during this study. The most abundant rock types in the study area are shallow marine mudstone and volcanic sandstone with subordinate limestone. These rocks are at least 2 km thick and grade locally into andesitic tuff. In most areas, the metasedimentary rocks are unconformably overlain by felsic to intermediate pyroclastic rocks and lava flows with exposed thicknesses of more than 1 km. Stratigraphic relationships between the various fault-bounded sequences of unfossiliferous rocks are not clear, but tentative lithologic correlations are shown in Figure 3.

The deformation of the metamorphic rocks in the Walker River drainage has not been studied in detail, but several generalizations about structure can nevertheless be made. Cleavage is well developed, especially in fine-grained metasedimentary rocks, but is less perfect in metavolcanic rocks. It appears to be axial planar to major folds. Where cleavage and bedding have been observed together, they are most commonly oriented at large angles to one another; this suggests that open rather than tight folds exist within the sequences. Unfortunately, marker beds that would allow recognition of large-scale folds have not been found. Stereographic plots of bedding attitudes within individual fault blocks suggest that major fold hinges are nearly horizontal and that their steeply dipping axial surfaces have trends averaging about N20°W. The Lobdell Lake sequence (Fig. 2) is unique in that the bedding and cleavage within it are consistently northeast trending and steeply dipping.

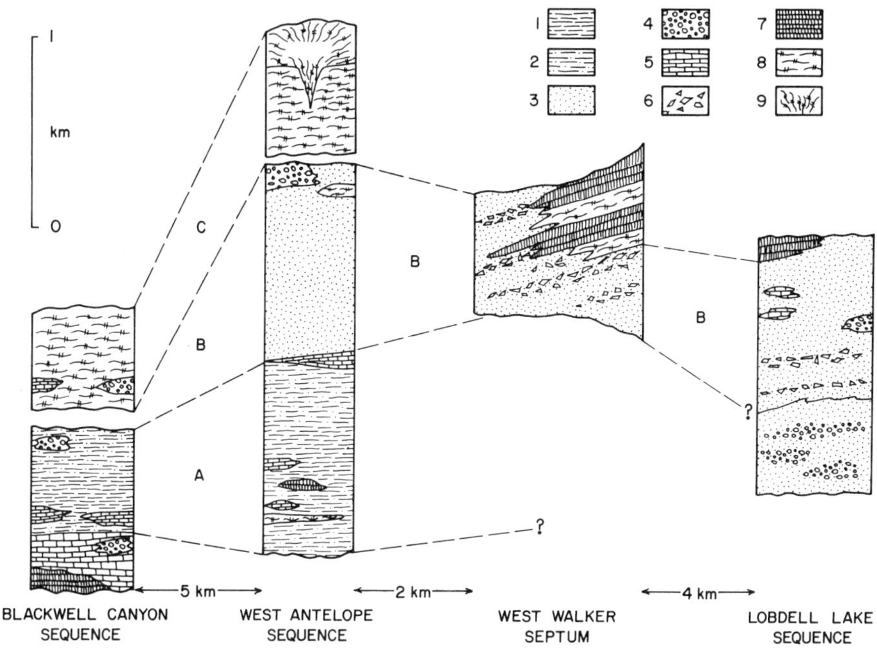

Figure 3. Preliminary stratigraphic correlations between metamorphic sequences. Unit B may be partly coeval with the West Walker pluton, and the upper part of unit C is roughly coeval with the China Garden pluton. Key to patterns (all shown in terms of premetamorphic lithologies): 1, shale; 2, siltstone; 3, lithic volcanic sandstone; 4, pebble and cobble conglomerate; 5, limestone; 6, tuff-breccia; 7, mafic lava; 8, rhyolite; 9, dacite.

Deformation appears to affect rocks in all sequences equally and therefore postdates the eruption of the youngest rocks (Fig. 3). Fold systems with similar orientations in other parts of the Sierra Nevada region are thought to have formed during Late Jurassic time (Kistler, 1966a; Schweickert and Cowan, 1975). If interpretations presented in this paper are accurate, the older plutons of the area must have experienced this same deformational event. It is notable that these Lower Jurassic (?) plutons do not display schistosity except in local marginal facies (discussed below). In fact, none of the dated Lower Jurassic and Upper Triassic plutons in regions south of the Walker River drainage have been reported to be schistose.

Metamorphism of the volcanic and sedimentary rocks has reached the hornblende-hornfels facies in the inner parts of contact aureoles around younger plutons and is in the albite-epidote-hornfels facies elsewhere. Original textures and structures are for the most part well preserved.

SCOPE OF THIS PAPER

Figure 2 shows the general arrangement of 12 plutonic masses in the drainage area of the West Walker River and along Swauger Creek. The plutons have a compositional range from olivine gabbro to biotite granite and range from 1.5 to more than 50 km^2 in areal extent. Some plutons may be parts of much larger intrusive masses that were separated into fragments by younger, crosscutting intrusions. Petrographic and field data for the cartographic units are summarized in Table 1.

Evidence indicating shallow levels of emplacement of four of the older (Lower Jurassic?) plutons is presented here, and the regional significance of these plutons and their coeval volcanic rocks is evaluated in terms of the tectonic history of the Sierra Nevada region.

Older Felsic Plutons

The older felsic plutons in the drainage area of the West Walker River are interpreted to be of Early Jurassic age and are composed predominantly of quartz monzonite. Unlike the younger plutons that cut them, the older plutons have been extensively altered. They also characteristically possess chilled margins and locally developed hypabyssal marginal facies. Two of the older felsic bodies, the China Garden pluton and the Desert Creek pluton, are discussed below.

CHINA GARDEN PLUTON

The elliptical China Garden pluton (Fig. 4) underlies about 25 km^2. Seriate quartz monzonite of the pluton is distinctive, both in outcrop and in thin section. Cloudy, subhedral oligoclase laths as much as 5 mm long are surrounded and occasionally are replaced by smaller grains of anhedral quartz and K-feldspar. The plagioclase tablets contain delicate oscillatory zones and often occur in glomeroporphyritic clots; some large twinned grains may be synneusis twins (Vance, 1969). Anhedral biotite and hornblende occur in variable amounts, but biotite always exceeds hornblende in abundance.

All rocks of the China Garden pluton have undergone extensive subsolidus thermal histories and are severely altered. Textural relationships indicate that much K-feldspar formed by replacement of plagioclase and that large skeletal quartz grains grew by replacement of plagioclase and K-feldspar. Rocks at the present level of exposure evidently contained less quartz and K-feldspar originally. The interiors of surviving plagioclase crystals contain fine aggregates of any of the following minerals: sericite, clinozoisite, epidote, and biotite.

In all rocks, biotite and hornblende form ragged, interstitial clots. Hornblende is generally bluish green, is often fibrous, and may be actinolitic. Some may have replaced original clinopyroxene. Biotite often replaces hornblende along cleavages. In some rocks, hornblende and biotite are completely replaced by clots of chlorite, epidote, and magnetite. Small grains of ilmenite are generally embayed and rimmed by fine-grained sphene.

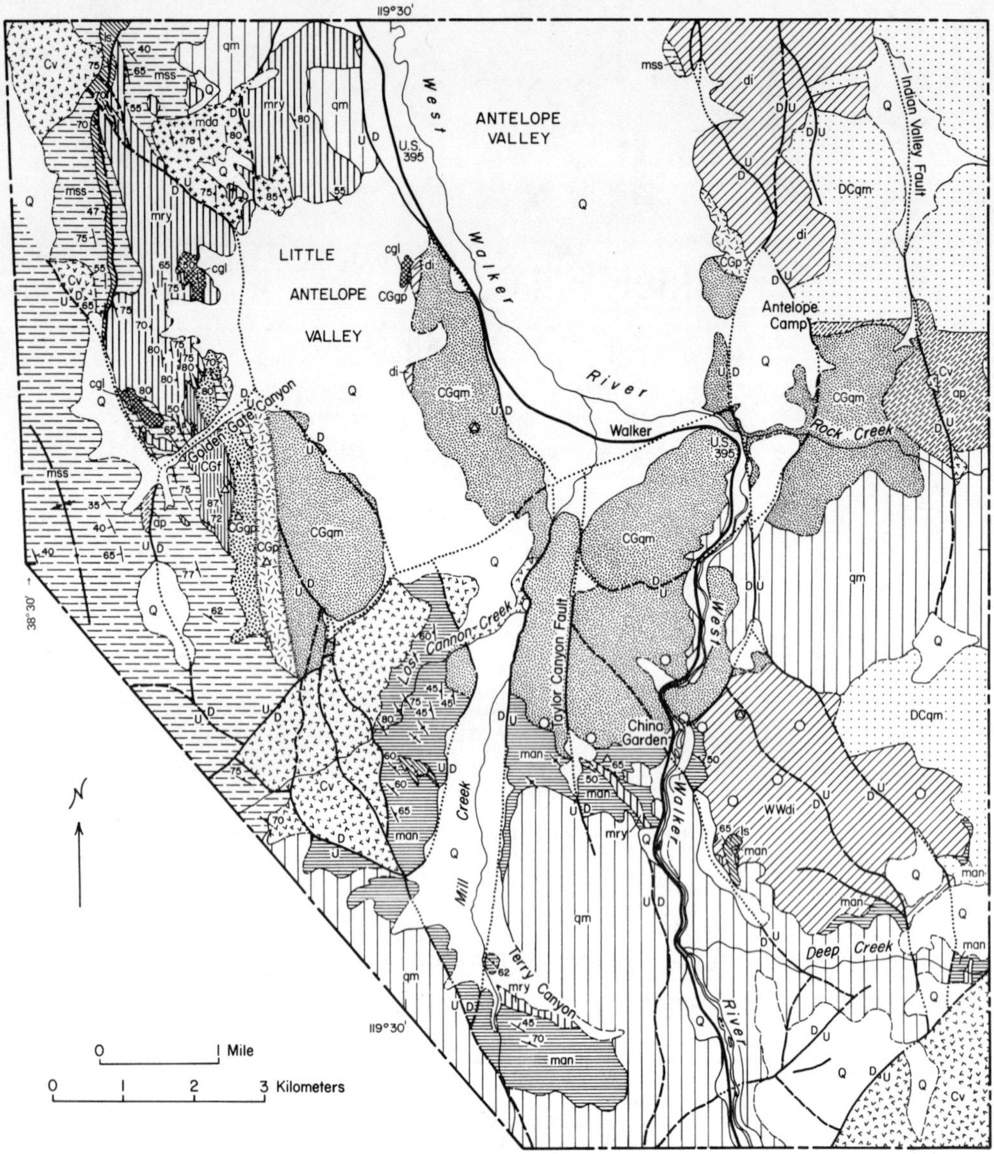

lithologic contact, approximate, concealed, gradational

fault, approximate, concealed

strike and dip of bedding; foliation or cleavage

modally analyzed sample; chemically analyzed sample

syncline axis

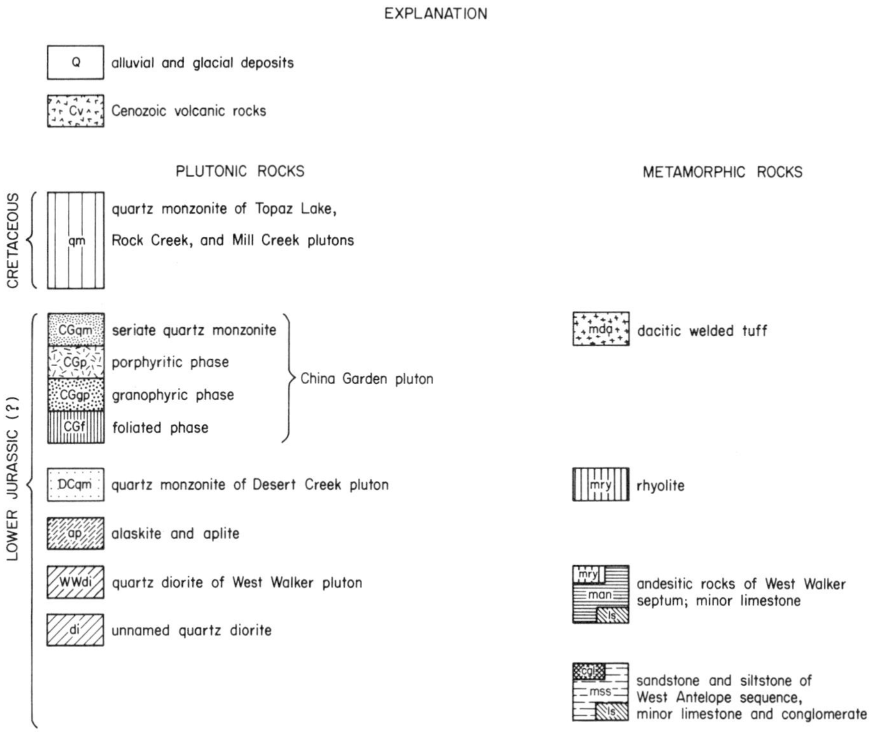

Figure 4. Geologic map (left; explanation, above) of the China Garden pluton and West Walker pluton.

Marginal units (discussed below) display all of the above alterations except the growth of large skeletal quartz crystals.

Modes and Chemistry

Six modes for the China Garden pluton are listed in Table 2; quartz, K-feldspar, and plagioclase, recalculated to 100 percent, are plotted as QKP percentages in Figure 5. In general, the rock is hornblende-biotite quartz monzonite, although some modes plot near the granodiorite-quartz monzonite boundary. Four chemical analyses of samples from this pluton are listed in Table 3, and sample localities are plotted in Figure 4. Sample 65A, with notably less SiO_2 than the others, comes from an area choked with metasedimentary inclusions. Sample 218 is seriate quartz monzonite typical of all five main parts of the pluton. Samples 196 and 195 (granophyric quartz monzonite porphyry and foliated quartz monzonite, respectively)

TABLE 2. MODES

Pluton and sample no.		Q (%)	K (%)	P (%)	Hbl (%)	Bt (%)	Cpx (%)	Opaque (%)	C. I.	Q* (%)	K* (%)	P* (%)
China Garden	218	22.7	27.4	39.5	2.5	5.7	0.0	2.0	10.5	25.3	30.6	44.1
	5	20.4	33.7	37.2	3.8	2.5	0.0	1.0	8.5	21.0	36.2	42.8
	23	18.0	29.4	41.0	1.2	4.5	0.0	2.2	11.5	20.4	33.3	46.4
	9	17.6	28.2	46.2	1.3	5.5	0.0	0.3	7.8	19.2	30.6	50.2
	90A	22.2	26.4	44.4	3.0	2.3	0.0	1.2	6.5	23.9	28.4	47.7
	176	17.0	20.8	51.8	0.5	8.5	0.0	0.8	10.3	18.9	23.2	57.9
Average of 6		19.6	27.8	43.4	2.1	4.8	0.0	1.2	9.2	21.6	30.6	47.8
Desert Creek	EWH	20.4	36.5	36.4	4.3	0.0	0.8	0.0	6.7	21.8	39.1	39.0
	13	18.5	35.4	35.5	1.8	0.0	7.0	0.5	10.6	20.7	39.6	39.7
	41	15.5	36.9	33.9	10.3	0.0	0.0	1.0	13.7	17.9	42.8	39.3
	44B	19.0	22.5	45.0	0.0	0.0	11.5	0.0	13.5	22.0	26.0	52.0
	241	14.5	34.0	34.0	12.8	1.8	0.0	0.8	17.3	17.6	41.2	41.2
	244	28.7	24.7	35.2	3.5	5.0	0.0	1.7	11.5	32.4	27.9	39.7
	247.1	17.7	37.9	28.5	11.5	0.0	0.0	0.5	16.3	21.1	45.0	33.9
Average of 7		19.2	32.6	35.6	6.3		3.9	0.5	12.8	21.9	37.4	40.7
West Walker	12A	13.7	3.5	56.0	7.7	16.0	0.0	2.7	27.2	18.7	4.8	76.6
	14	12.0	10.7	50.0	14.0	10.8	0.0	1.7	27.3	16.5	14.7	68.8
	18	11.8	9.8	46.1	9.8	19.3	0.0	1.5	32.2	17.8	14.4	68.1
	19	8.0	2.5	61.4	14.7	10.5	0.0	2.0	28.0	11.1	3.5	85.5
	24B	11.0	3.3	49.9	10.8	22.0	0.0	2.7	35.9	17.1	5.1	77.7
	183	0.8	0.0	55.8	30.3	10.0	0.0	2.8	43.3	1.5	0.0	98.5
Average of 6		9.5	5.0	53.5	14.5	14.8	0.0	2.2	32.4	14.0	7.4	78.6

Note: Thin-section modes, average of 600 points counted. Q, quartz; K, K-feldspar; P, plagioclase; Hbl, hornblende; Bt, biotite; Cpx, clinopyroxene; Opaque, opaque minerals; C. I., color index.

*Values after recalculating so that (Q + K + P) = 100%.

TABLE 3. CHEMICAL ANALYSES

	218 (%)	65A (%)	196 (%)	195 (%)	A (%)	241 (%)	19 (%)	117 (%)	B (%)	SC41 (%)	SC36B (%)	C (%)
SiO_2	67.48	63.80	69.38	66.95	65.88	64.33	54.89	59.28	54.20	53.71	44.24	43.84
Al_2O_3	17.13	17.20	16.20	17.95	15.07	15.85	15.68	19.48	17.17	18.46	16.07	13.46
Fe_2O_3	0.75	1.46	0.72	1.18	1.74	1.23	4.83	2.26	3.48	0.27	0.64	2.20
FeO	1.72	2.39	1.79	2.01	2.73	2.46	4.73	3.80	5.49	5.68	9.62	9.24
MgO	0.88	1.58	0.79	1.07	1.38	2.42	3.74	2.84	4.36	4.70	17.63	19.91
CaO	2.08	3.58	1.69	2.08	3.36	3.32	7.16	6.70	7.92	10.87	7.68	8.10
Na_2O	4.55	4.73	3.71	3.22	3.53	3.80	3.75	1.32	3.67	3.71	1.48	1.33
K_2O	4.37	3.60	4.47	4.07	4.64	5.15	2.20	1.92	1.11	1.22	1.50	0.58
H_2O^+	0.55	0.66	0.50	0.71	0.52	0.80	1.24	1.07	0.86	1.22	1.50	0.50
H_2O^-	0.10	0.15	0.22	0.19	0.00	0.10	0.05	0.23	0.00	0.12	0.13	0.00
TiO_2	0.26	0.50	0.24	0.42	0.81	0.44	1.20	0.59	1.31	0.80	0.12	0.67
CO_2	0.00	0.00	0.00	0.00	0.00	0.00	0.00	0.00	0.00	0.00	0.00	0.00
P_2O_5	0.06	0.17	0.07	0.11	0.26	0.06	0.39	0.13	0.28	0.04	0.00	0.18
MnO	0.04	0.03	0.01	0.02	0.08	0.02	0.09	0.06	0.15	0.07	0.08	0.13
SO_3	0.09	0.00	0.00	0.00	0.00	0.11	0.00	0.00	0:00	0.12	0.10	0.00
Total	100.08	99.85	99.79	99.98		100.09	99.95	99.68		99.98	99.76	
Quartz	17.70	17.21	25.04	25.96		11.79	6.45	24.10		0.00	0.00	
Orthoclase	25.82	21.27	26.42	24.05		30.43	13.00	11.35		7.21	8.86	
Albite	36.50	39.99	31.40	27.20		32.10	31.80	31.20		31.50	5.90	
Anorthite	9.93	15.10	7.90	9.60		11.00	19.40	32.40		30.00	32.80	
Diopside	0.00	1.29	0.00	0.00		4.12	10.81	0.00		19.32	4.28	
Hypersthene	4.38	5.72	4.28	4.72		6.85	7.03	11.32		4.75	0.00	
Olivine	0.00	0.00	0.00	0.00		0.00	0.00	0.00		4.74	42.48	
% Fo in olivine	0.00	0.00	0.00	0.00		0.00	0.00	0.00		54.14	69.99	
Magnetite	1.09	2.12	1.04	1.71		1.78	7.00	3.28		0.39	0.93	
Ilmenite	0.49	0.95	0.46	0.80		0.84	2.28	1.12		1.52	0.23	
Apatite	0.14	0.40	0.16	0.26		0.14	0.91	0.30		0.09	0.00	
Corundum	1.28	0.00	2.35	4.73		0.00	0.00	3.36		0.00	0.00	

Note: Analyst Alberto Obregón, Instituto de Geologia Laboratorios de Química, Universidad Nacional Autónoma de México. Columns are 218—Quartz monzonite of China Garden pluton; lat 38°30.5′N, long 119°29.3′W. 65A—Quartz monzonite of China Garden pluton; lat 38°30′N, long 119°29.3′W. 196—Granophyric quartz monzonite porphyry of China Garden pluton; lat 38°30.3′N, long 119°31.4′W. 195—Foliated quartz monzonite of China Garden pluton; lat 38°30.5′N, long 119°31.6′W. A—Average hornblende-biotite adamellite (Nockolds, 1954). 241—Quartz monzonite of Desert Creek pluton; lat 38°28′N, long 119°24.5′W. 19—Quartz diorite of West Walker pluton; lat 38°28.5′N, long 119°26.8′W. 117—Meta-andesite of West Walker septum; lat 38°28.4′N, long 119°28′W. B—Average andesite (Nockolds, 1954). SC41—Hornblende gabbro of Swauger Creek complex; lat 38°22′N, long 119°19′W. SC36B—Olivine gabbro of Swauger Creek complex; lat 38°22′N, long 119°18.8′W. C—Average troctolite (Nockolds, 1954).

are chemically similar to the seriate quartz monzonite. All are somewhat richer in Al_2O_3 and poorer in total Fe than average adamellite (Nockolds, 1954). The CaO/SiO_2 and K_2O/SiO_2 ratios are similar to those reported for Lower Jurassic granitic rocks of the White Mountains by Bateman and Dodge (1970, Figs. 4 and 5), but are distinct from the ratios observed in younger Sierra Nevada sequences (Fig. 13).

Structures

Foliation and mineral lineation are notably absent in quartz monzonite of the interior of the pluton. Rocks in the core of the body are remarkably homogeneous, and samples from the various fault blocks are virtually indistinguishable. In contrast, the rocks of the pluton are finer grained and rather heterogeneous near its walls and in the marginal facies (discussed below), where foliated rocks are common.

Inclusions

Small, dark inclusions are irregularly distributed in the pluton but are concentrated near its margins especially along the western part. Most inclusions appear to be fragments of metasedimentary rocks from the walls of the pluton, but swarms of inclusions and large blocks of diorite occur on the east edge of Little Antelope Valley and near the mouth of Golden Gate Canyon (Fig. 4). Most inclusions are less than 10 cm across, but one enormous block, 1 km × 150 m, probably a small roof pendant, occurs at the mouth of Golden Gate Canyon.

Marginal Facies

The western part of the pluton contains several distinct mappable phases, two of which are not developed in other parts of the pluton. Variations in these units are great, and their mutual relationships are complex. In view of their probable bearing on the level of emplacement of the China Garden pluton, the units of the western edge of the pluton are discussed in detail below.

West of Little Antelope Valley, seriate granular quartz monzonite of the interior of the pluton grades westward into porphyritic quartz monzonite (Fig. 4; map unit CGp). Plagioclase phenocrysts are unchanged, but interstitial quartz and K-feldspar are finer grained. Similar porphyritic quartz monzonite was mapped along the northeastern edge of the pluton (Fig. 4) and occurs to a lesser extent along the chilled margin elsewhere.

Farther west, the porphyritic phase passes into a granophyric phase (Fig. 4; map unit CGgp), in which the plagioclase phenocrysts are fringed with granophyric intergrowths of quartz and K-feldspar that merge to form the groundmass of the rock. Tiny miarolytic cavities are conspicuous in this granophyric quartz monzonite porphyry. One-half kilometre north of Golden Gate Canyon, sills of granophyric quartz monzonite porphyry 5 to 50 m thick penetrate as far as 600 m into preintrusive metarhyolite (Fig. 4).

The zone of granophyric quartz monzonite porphyry averages about 400

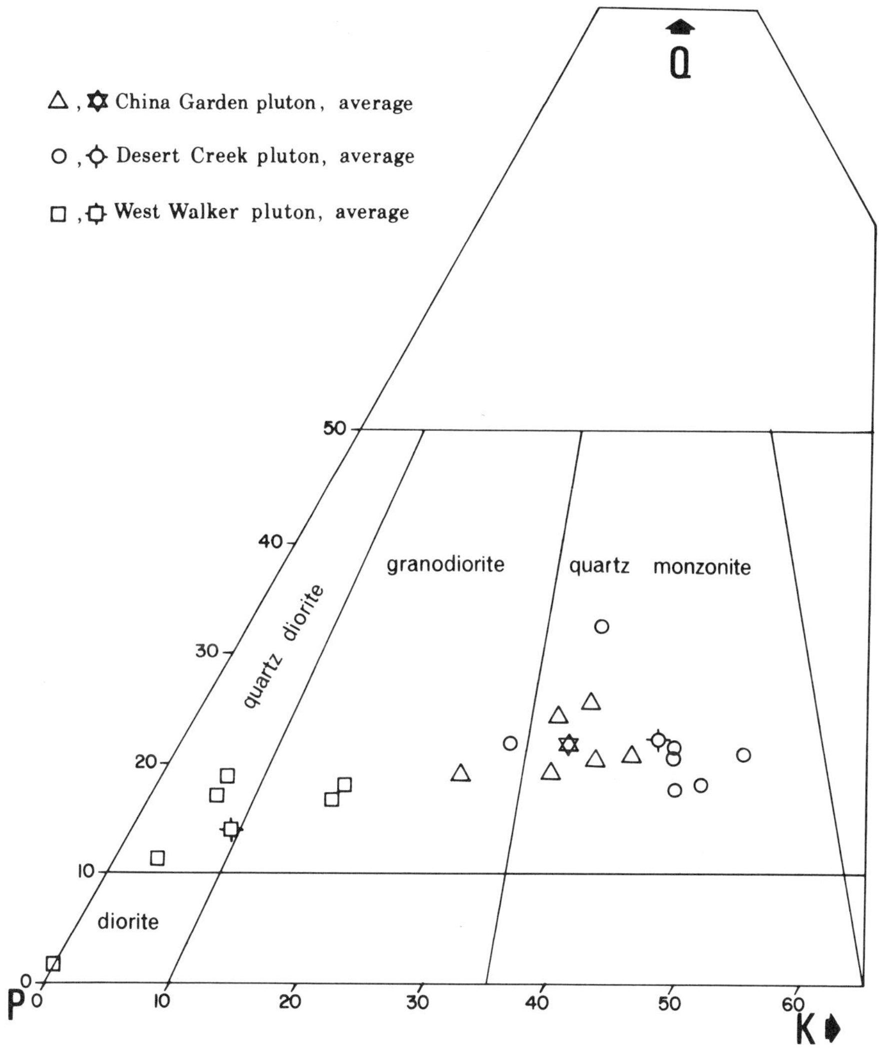

Figure 5. Plot of modes of representative samples. Quartz, K-feldspar, and plagioclase have been recalculated to 100 percent and are expressed as QKP percentages.

m across. At its western edge, weakly foliated granophyric quartz monzonite porphyry grades across 10 to 15 m into strongly foliated, schistose quartz monzonite (Fig. 4; map unit CGf), with abundant plagioclase augen visible in hand specimen. This unit is heterogeneous, contains abundant flattened dark xenoliths, and averages about 300 m across. A thin selvage of granophyric quartz monzonite porphyry separates the foliated rocks from the wall rocks, which are disrupted by irregular dikes and protrusions of the porphyry for as much as 10 m from the main contact.

In each of these marginal units, plagioclase tablets are morphologically and compositionally identical to those occurring in the core of the pluton, and here they form both individual subhedra and glomeroporphyritic clots. Broken plagioclase crystals, though not abundant, are more conspicuous in the foliated quartz monzonite, and some appear rounded or abraded.

Thin sections of foliated quartz monzonite cut normal to the foliation show that the plagioclase crystals are flanked by coarse, lensoid aggregates of quartz that parallel the foliation. Wisps and streaks of brown and reddish-brown biotite together with the elongate lenses of quartz define the foliation, but plagioclase phenocrysts or augen lie at all angles to it.

The foliated quartz monzonite clearly shows evidence of flowage that probably occurred while the glassy groundmass was still plastic. Groundmass devitrification or crystallization probably accompanied the flowage, although the biotite growth was clearly mimetic after the original flowage foliation.

Gradational contacts between the various marginal units are subvertical and are subparallel both to the intrusive contact and to the steep foliation of the granophyric quartz monzonite porphyry and the foliated quartz monzonite.

Metasedimentary inclusions are common to all of the marginal units described above, but their abundance is greatly variable. They appear to be most common near the wall rocks on the west and in the vicinity of the large metasedimentary pendant at the mouth of Golden Gate Canyon. All are strongly recrystallized and contain metamorphic mineral assemblages indicative of the hornblende-hornfels facies. In addition, the foliated quartz monzonite contains flattened mafic xenoliths that are not found in the other units. These are generally biotite-rich masses whose textures resemble the texture of the foliated quartz monzonite itself.

Interpretation of Marginal Units

It is possible that the complex marginal units of the pluton represent a protoclastic border, formed as the pluton, largely crystalline, bulged its walls as it ascended. However, the pluton–wall rock contact is sharp and irregular, with a thin veneer of unfoliated granophyric quartz monzonite porphyry adjacent to the contact, and exhibits no evidence of shearing.

There is no evidence that nearly solid magma abraded and smoothed its walls to any degree during its ascent. The geometry of the contact indicates instead that blocks were continually wedged apart and split off from the walls and were incorporated piecemeal into the partly crystalline magma.

Bedding and cleavage of wall rocks are subparallel to the western margin of the pluton. This suggests that the foliated quartz monzonite formed during the deformation of the wall rocks. However, the facts that (1) the foliated quartz monzonite is a localized unit and does not form an annulus around the pluton and (2) a thin veneer of unfoliated granophyric quartz monzonite porphyry is developed adjacent to the western contact both argue that the foliation in this unit is not a result of the deformation that affected the wall rocks of the pluton. Some small-scale disruption of bedding occurs as much as a few metres from the contact, but there is no clear evidence that the pluton bulged aside its walls as it ascended. Finally, granophyric quartz monzonite is difficult to explain as a protoclastic border unit.

I suggest, instead, that the border phases of the pluton are part of a feeder system or conduit that had access to the surface in the form of an eruptive vent. The foliated quartz monzonite would thus record late magmatic viscous flow and crystallization or devitrification of glassy crystal-rich magma as it approached a constricted vent above. The granophyric quartz monzonite porphyry might be interpreted as quenched crystal-rich magma that did not experience appreciable viscous flow. Granophyric textures have been produced experimentally by devitrification of natural glasses at elevated temperatures (W. C. Luth, 1970, oral commun.). In addition, Harker (1904), studying epizonal rocks on the Isle of Skye, observed gradations between marginal plutonic rocks with granophyric textures and glassy rocks with spherulitic structures and concluded that these features represented different stages of devitrification of rocks that were originally glassy. One might further argue that the entire marginal facies of the China Garden pluton was originally quenched, perhaps by pressure release as access to the surface was gained, and that the foliated quartz monzonite originally consisted of crystals and glass that deformed plastically as it flowed. Some samples of foliated quartz monzonite contain imperfect relicts of granophyre; this suggests that granophyric quartz-feldspar intergrowths that may have originally existed in this unit were largely destroyed during viscous flow.

The variations in groundmass textures can be explained as follows: In the core of the pluton, crystal-rich magma with viscosity low enough for plagioclase crystals to collect into glomeroporphyritic clots crystallized slowly and thus acquired seriate textures. The crystal-rich liquid near the western margin and near contacts with wall rocks elsewhere cooled more quickly and produced porphyritic quartz monzonite with a quartz–K-feldspar mesostasis. The part of the magma that was pressure-quenched at the western margin developed a crystal-rich glass that eventually devitrified. Where the viscous, crystal-charged glass flowed upward, either the early formed granophyric textures were disrupted and granulated, or the glassy material was attenuated and the microlites that had begun to appear were granulated. The result is a groundmass of granular quartz, K-feldspar grains, and elongate trains of quartz.

Possible Extrusive Equivalent

The above interpretation implies that the foliated quartz monzonite represents the feeder zone for a dikelike conduit through which viscous material was routed toward the surface where pyroclastic rocks were erupted. North of Little Antelope Valley is a metamorphosed pyroclastic unit, with an areal extent of about 2 km^2, that is demonstrably younger than the metarhyolite intruded by the China Garden pluton (Fig. 4). This unit may be the pyroclastic equivalent of the foliated quartz monzonite of the China Garden pluton. The rock is mainly crystal-rich, dacitic welded tuff and is characterized by irregular, broken plagioclase crystal fragments embedded in a very fine grained groundmass of quartz and K-feldspar. Flattened, felted lithic fragments and lensoid quartz segregations, together with metamorphic biotite, define the foliation or eutaxitic structure.

Several lithologic and mineralogic similarities between the foliated quartz monzonite and the dacitic welded tuff support the hypothesis of their equivalence. The foliated quartz monzonite contains about 20 to 25 percent plagioclase crystals—some in clots and a few of them broken. The welded tuff contains about 15 to 20 percent plagioclase crystals—nearly all of them angular fragments but, significantly, with occasional clots of crystals still recognizable. Dark, compressed xenoliths are common in the foliated quartz monzonite and are also found in the welded tuff. Pleochroic dark brown allanite, a common accessory mineral in the China Garden pluton, occurs also in both the foliated quartz monzonite and the dacitic welded tuff. The only other known occurrence of allanite in the area is in the Desert Creek pluton (described below).

Characteristics of the China Garden pluton that are common to two Tertiary epizonal plutons in the Northern Cascade Mountains are summarized in Table 4. These Tertiary plutons are much more extensively altered than Cretaceous plutons near them. Fiske and others (1963) concluded that rocks of the Tatoosh pluton were autometamorphosed by volatile components released upward during explosive eruptions. The replacement textures noted in the China Garden pluton are generally similar to alteration in the Cloudy Pass batholith described by Cater (1969) and in the Tatoosh pluton described by Fiske and others (1963); such textures possibly resulted

TABLE 4. CHARACTERISTICS OF SELECTED EPIZONAL PLUTONS

Characteristic compared	Cloudy Pass batholith (Cater, 1969)	Tatoosh pluton (Fiske and others, 1963)	China Garden pluton (this paper)
Chief rock type	Granodiorite	Granodiorite	Quartz monzonite
Oscillatory zoned euhedral plagioclase	Common	Common	Common
Interstitial K-feldspar and quartz	Present	Present	Present
Plagioclase replaced by late K-feldspar and quartz	Yes	Not stated	Yes
Rocks altered	Yes	Yes	Yes
Flow banding and mineral lineation	Present locally	Absent	Absent
Intrusive breccias	Present	Present	Present?
Discordantly intrudes wall rocks	Yes	Yes	Yes
Chilled borders	Present	Present	Present
Porphyritic facies	Present	Present	Present
Granophyre dikes or sills with miarolytic cavities near top	Present	Present	Present
Flow-banded hypabyssal marginal facies	Present	Present	Present
Hypabyssal or extrusive dacite porphyry or welded tuff adjacent to pluton	Present	Present	Present

from the introduction of K_2O and SiO_2 from deeper levels of the pluton by volatile components that accumulated near its roof. Some of the alteration of China Garden rocks, however, probably is a result of contact metamorphism by younger plutons.

DESERT CREEK PLUTON

The Desert Creek pluton, more than 60 km^2 in area, lies east of Antelope Valley (Fig. 6) and has many characteristics in common with the China Garden pluton. Coarse-grained seriate rocks of the pluton contain 4- to 10-mm-long K-feldspar subhedra and anhedra, in addition to subhedral 4- to 5-mm-long plagioclase tablets that (as in the China Garden pluton)

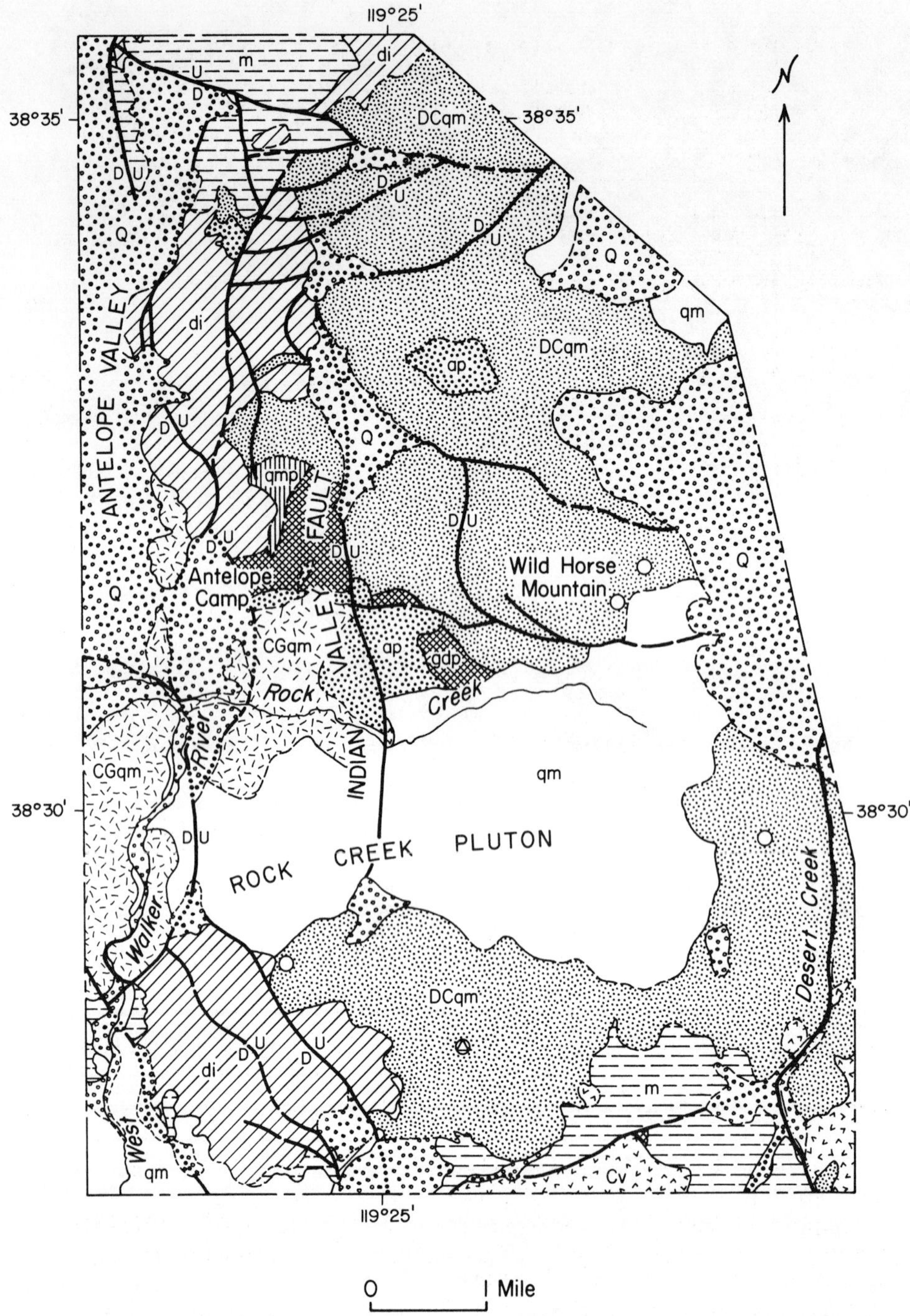

119°25'
38°35'
38°30'
N
ANTELOPE VALLEY
INDIAN VALLEY FAULT
Antelope Camp
Wild Horse Mountain
ROCK CREEK PLUTON
Rock Creek
Walker River
West Walker
Desert Creek
Q
m
di
DCqm
qm
ap
qmpl
CGqm
gdp
Cv
D
U
0 1 Mile
0 1 2 3 Kilometers

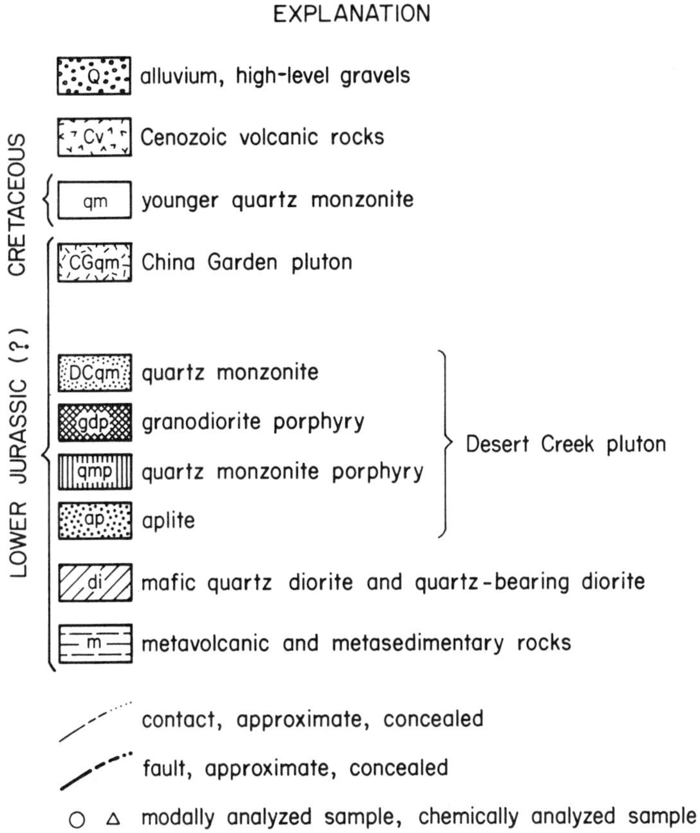

Figure 6. Geologic map (left; explanation, above) of Desert Creek pluton and Rock Creek pluton.

frequently form glomeroporphyritic clots and synneusis twins. Plagioclase is normally zoned from an average of An_{36} in cores to An_{20} in rims. Most samples contain moderate to abundant clinopyroxene, often largely replaced by bluish-green hornblende. The low original content of quartz that is evident in thin sections and modes (Table 2) and the abundance of clinopyroxene give the rock direct monzonitic affinities. Similar pyroxene-bearing monzonites and other alkali-calcic rocks occur in the Inyo Mountains, 150 km to the southeast (Sylvester and others, 1969; Fig. 11), and these have yielded Early Jurassic radiometric ages (McKee and Nash, 1967).

Rocks of the Desert Creek pluton, like those of the China Garden pluton, show textural evidence of replacement and alteration. Skeletal quartz grains are again abundant, and plagioclase crystals have been replaced by K-feldspar around their margins. The large K-feldspar anhedra evidently grew by replacement, and if they were removed, these rocks would closely resemble those of the China Garden pluton. Plagioclase is dusted with fine inclusions of clinozoisite, sericite, and epidote. Pale bluish-green hornblende with relict cores of clinopyroxene generally forms ragged grains that are associated with clots of sphene and magnetite. Biotite replacing hornblende was observed in the part of the pluton south of the Rock Creek pluton (Fig. 6).

Modes and Chemistry

The interior of the Desert Creek pluton is fairly homogeneous, with all sample modes plotting in the quartz monzonite field. These rocks characteristically contain more than 37 percent K-feldspar on a QKP plot (Fig. 5). One chemical analysis for this pluton is given in Table 3. Although containing less SiO_2 than China Garden rocks, the Desert Creek sample is notably richer in K_2O. Na_2O, CaO, and total Fe oxides are not appreciably different from analyzed China Garden samples.

Structures

The interior of the Desert Creek pluton, like that of the China Garden pluton, lacks megascopic structures. An irregular chilled margin, averaging 5 to 10 m thick, is developed near contacts with older rocks. Porphyritic quartz monzonite along the margin resembles rocks in the interior of the pluton and possesses identical K-feldspar and andesine subhedra. As in chilled zones of the China Garden pluton, the border rocks have a fine-grained granular groundmass. Contacts with wall rocks appear sharp but are poorly exposed. No inclusions were noted in the Desert Creek pluton.

Marginal Facies

Unlike the homogeneous interior of the pluton, the west flank of the Desert Creek pluton adjacent to the China Garden pluton is made up of a heterogeneous assemblage of porphyries, alaskites, aplites, and occasional granophyres. Exposures are poor, and contacts between subunits cannot be located accurately. Most porphyries contain subhedral K-feldspar and plagioclase phenocrysts like those in the quartz monzonite of the interior of the pluton, but generally the porphyries are more mafic. Abundant brown biotite forms knots in the groundmass, along with less abundant bluish-green hornblende and infrequent clinopyroxene relicts. Modes of the porphyritic rocks range from granodiorite to granite, but granodiorite and quartz monzonite are most common. Some granodiorite porphyries occur as dikelike masses containing appreciable clinopyroxene but lacking biotite and hornblende. Specks of brown and green allanite were recognized in most thin sections.

Age relationships between the Desert Creek and China Garden plutons are obscured by the complex buffer zone of porphyritic and alaskitic rocks between them. However, dikes of China Garden quartz monzonite were observed intruding granodiorite porphyry believed to be part of the Desert Creek pluton just east of the Indian Valley fault (Fig. 6). If the granodiorite porphyry was emplaced at the same general time as the Desert Creek pluton, the China Garden is the younger of the two plutons.

Interpretation

Textures, chilled margins, and hypabyssal border facies of the Desert Creek pluton are analogous to those of the China Garden pluton. Although the porphyries at the west edge of the pluton are not clearly understood, they may represent a series of subvolcanic hypabyssal feeders, as suggested for marginal units at the west edge of the China Garden pluton.

OTHER "OLD" QUARTZ MONZONITE PLUTONS

North and east of Swauger Creek (Fig. 2) lies the East Fork pluton, the oldest silicic pluton in the Swauger Creek drainage. Though only studied in reconnaissance, it is apparent that rocks of this body have textures, compositions, and alterations similar to those of rocks of the Desert Creek pluton. The East Fork pluton underlies about 10 km^2 in the area mapped and is known from reconnaissance to extend several kilometres farther east toward the East Walker River. Rocks indistinguishable from the East Fork quartz monzonite also occur near the confluence of Buckeye and Eagle Creeks, 15 km south of the East Fork of Swauger Creek (Fig. 10). The older, altered granitic rocks described in the preceding sections thus are not isolated or unusual occurrences but appear representative of a substantial terrane underlain by Lower Jurassic plutons.

Older Mafic Plutons

The four small mafic plutons that lie in the Sweetwater Range between the east and west forks of the Walker River predate and are intruded by all felsic plutons that surround them. The mafic plutons have been contact metamorphosed, and their distribution is closely related to remnants of metavolcanic rocks. Compositions range from mafic quartz diorite to olivine gabbro or troctolite. The West Walker pluton and the Swauger Creek pluton are discussed below.

WEST WALKER PLUTON AND RELATED ROCKS OF WEST WALKER SEPTUM

The 7-km^2 West Walker pluton and the adjacent metamorphosed andesitic rocks of the West Walker septum lie along the southern boundary of the China Garden pluton (Fig. 4). This pluton is interpreted as a slightly older and intermediate analogue of the shallow-level felsic plutons discussed above.

The quartz diorite of the pluton seems abnormally dark in hand specimen, but its color index is only about 25 to 45 (Tables 1 and 2). Most rocks display cumulate textures, with euhedral to subhedral plagioclase laths as much as 1 cm long composing about 50 to 60 percent of the rock. Plagioclase compositions range from An_{40} to An_{50}, samples low on the east wall of the West Walker Canyon containing progressively more calcic plagioclase than samples from higher elevations. The correlation between the percentage of An and sample elevation is shown in Figure 7 and is reminiscent of similar correlations noted in large layered intrusions (Wager and Brown, 1967, Fig. 16, p. 34). Sodic labradorite occurs at the lowest exposed levels, and here the rock could be termed gabbro. Otherwise, by most criteria, the rock is quartz diorite (see Modes and Chemistry, below).

K-feldspar, quartz, hornblende, and biotite occupy interstices between plagioclase laths. K-feldspar and quartz form anhedral xenoblastic grains and make up about 15 percent of the rock. Mafic minerals clearly do not represent primary precipitates from a melt. Hornblende generally forms

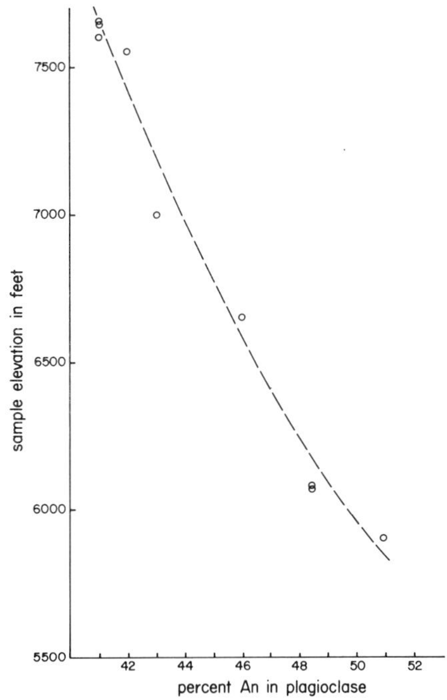

Figure 7. Plot of plagioclase composition against sample elevation in the West Walker pluton.

bluish-green clusters of fibrous crystals and locally contains clinopyroxene relicts. Some hornblende crystals resemble metamorphic porphyroblasts. Tiny, deep olive-green biotite flakes commonly occur within ragged hornblende crystals, together with abundant magnetite. Tiny needles and plates of both hornblende and biotite have been observed growing into and within plagioclase crystals.

Plagioclase is dusted with a very fine unresolvable mineral, and this accounts for the dark appearance of most samples. Recrystallization has occurred near points of contact between adjacent plagioclase crystals. Subsolidus replacement textures, though present, are not as convincing as in the China Garden and Desert Creek plutons.

Modes and Chemistry

Six modes for the West Walker pluton are listed in Table 2. The modes average about 10 percent quartz but contain 25 to 45 percent mafic minerals. Thus, most of the samples plot well within the quartz diorite field (Fig. 5) when quartz, K-feldspar, and plagioclase are recalculated to 100 percent and plotted as QKP percentages. Therefore, the dominant rocks with slightly more than 10 percent quartz might be called mafic quartz diorite and others, with less than 10 percent modal quartz, quartz-bearing diorite. A chemical analysis of sample 19 is listed in Table 3, together with several published analyses of andesitic and related rocks. Sample 19 compares quite closely with Nockolds's (1954) average monzonite (not shown in Table 3) in all oxides except K_2O, which is about 2.5 percent lower in sample 19. Al_2O_3 and MgO are slightly lower and K_2O and total Fe are higher in sample 19 when compared with Nockolds's average hornblende-biotite diorite. Except for CaO (which is low) and K_2O (which is high), sample 19 appears quite similar to Nockolds's (1954) average andesite and to Waters's (1962) average Cascade andesite, but is rather dissimilar to Chayes's (1969) average U.S. Cenozoic andesite in several respects. Indeed, on the basis of the single analysis, mafic quartz diorite of the West Walker pluton is difficult to pigeonhole. In general, if the notably high K_2O level is overlooked, sample 19 is chemically similar to andesite. It is intriguing to note that there is evidence that other mafic plutons of the eastern Sierra Nevada may also possess higher K_2O contents for similar SiO_2 percentages than do mafic intrusions along the western flank of the Sierra Nevada batholith (compare Bateman and others, 1963, p. 29, and Best, 1963), but chemical data on these eastern mafic bodies are still very sparse.

Structures

Careful inspection of most outcrops of the pluton reveals the presence of a weak subhorizontal foliation that is defined by statistical orientation of (010) faces of euhedral to subhedral plagioclase tablets. Mafic minerals do not contribute to the foliation, and megascopic layering defined by alternations of dark and light minerals is not present. Owing to the gentle dip of the foliation, measurements are only approximate; these indicate a westward dip of 0° to 10°.

Marginal Variation

Rather coarse-grained throughout, the body appears remarkably homogeneous and free of inclusions. On its west edge near China Garden, however, an abrupt transition to fine-grained diorite occurs near the contact with metavolcanic rocks. The fine-grained diorite has a distinctly hypabyssal appearance and grades outward into typical extrusive volcanic rocks (described below).

Other Small Mafic Bodies

Two small dioritic plutons similar to the West Walker pluton occur east of Antelope Valley. Neither of these plutons was studied in detail. One, 6 km^2, extends several kilometres north from Blackwell Canyon (at north edge of map in Fig. 6) to Rickey and Risue Canyons on the north and east. The other, 8 km^2, lies north of Antelope Camp. Both plutons bear a striking resemblance to the West Walker pluton, both in field appearance and in textures and mineral content. However, the 8-km^2 body, unlike the West Walker pluton, may postdate the Desert Creek pluton, for some fine-grained marginal variants are developed adjacent to contacts with the Desert Creek pluton. In any case, age relationships between the Desert Creek pluton and this dioritic pluton are uncertain because of the intervening granodiorite and diorite porphyries between the two plutons, and the poor quality of exposures. As in the West Walker pluton, rocks of both these small quartz diorite plutons are typified by cumulus textures in which large subhedral andesine laths define a weak subhorizontal foliation and are surrounded by finer interstitial K-feldspar, quartz, hornblende, and biotite. All three dioritic plutons could be remnants of a formerly much larger mafic quartz diorite body, perhaps analogous to the intrusion in the West Humboldt Range (Speed, 1963), 100 km northeast, but there is no direct evidence for this.

Metamorphosed Volcanic and Hypabyssal Rocks of West Walker Septum

Dominantly andesitic rocks occupy a thin septum lying between the China Garden pluton and the younger Mill Creek pluton (Fig. 4). This septum extends 2.5 km west from the edge of the West Walker pluton to Mill Creek; rocks of this sequence also crop out in Terry Canyon and along the ridge between Lost Cannon and Mill Creeks. Wedged between the two plutons, most of the rocks in the septum possess metamorphic mineral assemblages indicative of the hornblende-hornfels facies. Both the China Garden pluton and the Mill Creek pluton sharply truncate bedding of lava flows and tuff in the septum.

Contact relationships with the West Walker pluton are more complex. At the west edge of the pluton, fine-grained mafic quartz diorite (described above) lies with sharp contact against coarser rocks characteristic of the interior of the pluton. In some areas, alternating zones of coarse- and fine-grained diorite cause the contact to appear gradational. A contact breccia with fragments of meta-andesite 10 cm to several metres in length is developed locally. Exposures are poor near the river, and units cannot be traced with certainty. West of the river, fine hypabyssal andesite or diorite indistinguishable from that near the pluton occurs in tabular, sill-like masses that apparently interleave with porphyritic andesite lava flows. Throughout this area, discontinuous blocks of extrusive rock appear to be embedded in the hypabyssal rock.

Andesitic flows, occasionally interbedded with more silicic rocks, increase in abundance and predominate farther west in the septum. Relict volcanic textures are everywhere preserved in these rocks. The formerly glassy groundmass of most rocks has recrystallized to a granoblastic aggregate of quartz and feldspar, especially near contacts with intrusive bodies. Pale watery green hornblende occurs in two habits: small poikiloblasts, with inclusions of plagioclase and opaque minerals, and stubby grains that are pseudomorphs after volcanic pyroxene. Oriented biotite flakes impart a faint to moderate schistosity to many rocks. The abundance of biotite in these rocks is a reflection of their relatively high K_2O content. A chemical analysis of one sample is listed in Table 3.

Metamorphosed tuff, tuff-breccia, and volcanic sandstone occur with lava flows west of the Taylor Canyon fault, and these form an important part of the sequence along the east side of Mill Creek and Terry Canyon. West of Mill Creek, coarse volcanic sandstone and conglomerate are more frequently interbedded with the tuff, and several discontinuous lenses of calcarenite and limestone occur. It is probable that this sequence grades westward into the dominantly sedimentary succession south of Golden Gate Canyon, but downfaulted Cenozoic volcanic cover along Lost Cannon Creek (Figs. 2 and 4) obscures this relationship. Most bedding attitudes in the septum strike northwest and dip northeast.

Rocks of the West Walker pluton and metavolcanic rocks in the West Walker septum are mineralogically similar, both in terms of original lithologies and metamorphic mineral assemblages. To investigate the chemical similarity between the mafic quartz diorite and meta-andesite, sample 19 (representative of mafic quartz diorite) and sample 117 (representative of meta-andesite) were selected for analysis (Table 3). As discussed above, sample 19 chemically resembles some andesitic rocks, but is high in K_2O. Except for a marked deficiency of Na_2O and a slight excess of Al_2O_3, sample 117 compares quite closely both with Chayes's (1969) average U.S. Cenozoic andesite and with several analyzed samples of historically erupted continental andesite of Trident and Redoubt volcanoes, Alaska Peninsula (Forbes and others, 1969, Tables 1 and 2). According to Forbes and others (1969), these highly siliceous continental andesitic rocks are chemically distinct from Aleutian (island arc) andesites that contain more Fe and less Ca and from Nockolds's (1954) average andesite that contains less SiO_2 and more Ca and total Fe. No ready explanation is apparent for the abnormally low Na_2O content of sample 117, but in most other respects, sample 117 chemically resembles continental andesite.

Considered together, samples 19 and 117 are not closely similar chemically; the analyzed meta-andesite is markedly higher in SiO_2 and Al_2O_3 and is correspondingly poorer in Na_2O, K_2O, and total Fe than is the mafic quartz diorite. These differences rule out a simple or direct relationship between the two rock types, but do not eliminate the possibility (discussed below) that similar andesitic lavas may have been erupted from the magma now represented by the mafic quartz diorites—especially since, if the crystal-accumulation hypothesis is correct for the West Walker pluton, samples of cumulate rocks would not be expected to approximate the composition of the magma or magmas from which they formed. Obviously, more analytical data are required to define patterns of variation within the septum and within the pluton, before detailed petrogenetic models can reasonably be formulated.

Interpretation of Relationships between West Walker Pluton and Septum

The West Walker pluton intruded a sequence of andesitic rocks whose chemical composition is somewhat similar to its own. Andesite and mafic quartz diorite sills and andesitic lava flows are found chiefly in the part of the septum nearest the pluton and decrease markedly in abundance away from the pluton. The complex intrusive and extrusive relationships discussed above could have originated if the West Walker pluton invaded its own volcanic cover produced by repeated eruptions of chemically similar andesitic lavas and tuffs. Sills and dikes of andesitic magma penetrated and bulged the volcanic wall rocks, especially near the surface. It is therefore suggested that the mafic quartz diorite crystallized and differentiated at a relatively shallow level in a subvolcanic magma chamber beneath its own outpourings of mafic to intermediate lavas.

The main part of the pluton formed from a fairly mafic magma in which early formed plagioclase crystals settled out, forcing the crystallization of progressively more sodic plagioclase from the melt. Only plagioclase appears to have settled out of the magma at these levels of exposure, although the occurrence of more calcic plagioclase at deeper levels may indicate that more refractory phases settled to depths below the present level of exposure.

SWAUGER CREEK MAFIC COMPLEX

The Swauger Creek mafic complex is a layered, sill-like intrusion and may be a mafic analogue to the silicic and intermediate bodies described above. This 4.5-km^2 complex occupies the southern part of an erosional window through Cenozoic volcanic cover in the southeast part of the study area (Figs. 2 and 8).

Remnants of Cenozoic volcanic cover scattered along the east edge of Swauger Creek Canyon record westward tilting of about 10° to 20°. The mappable units of the complex define a layered, sill-like body whose floor

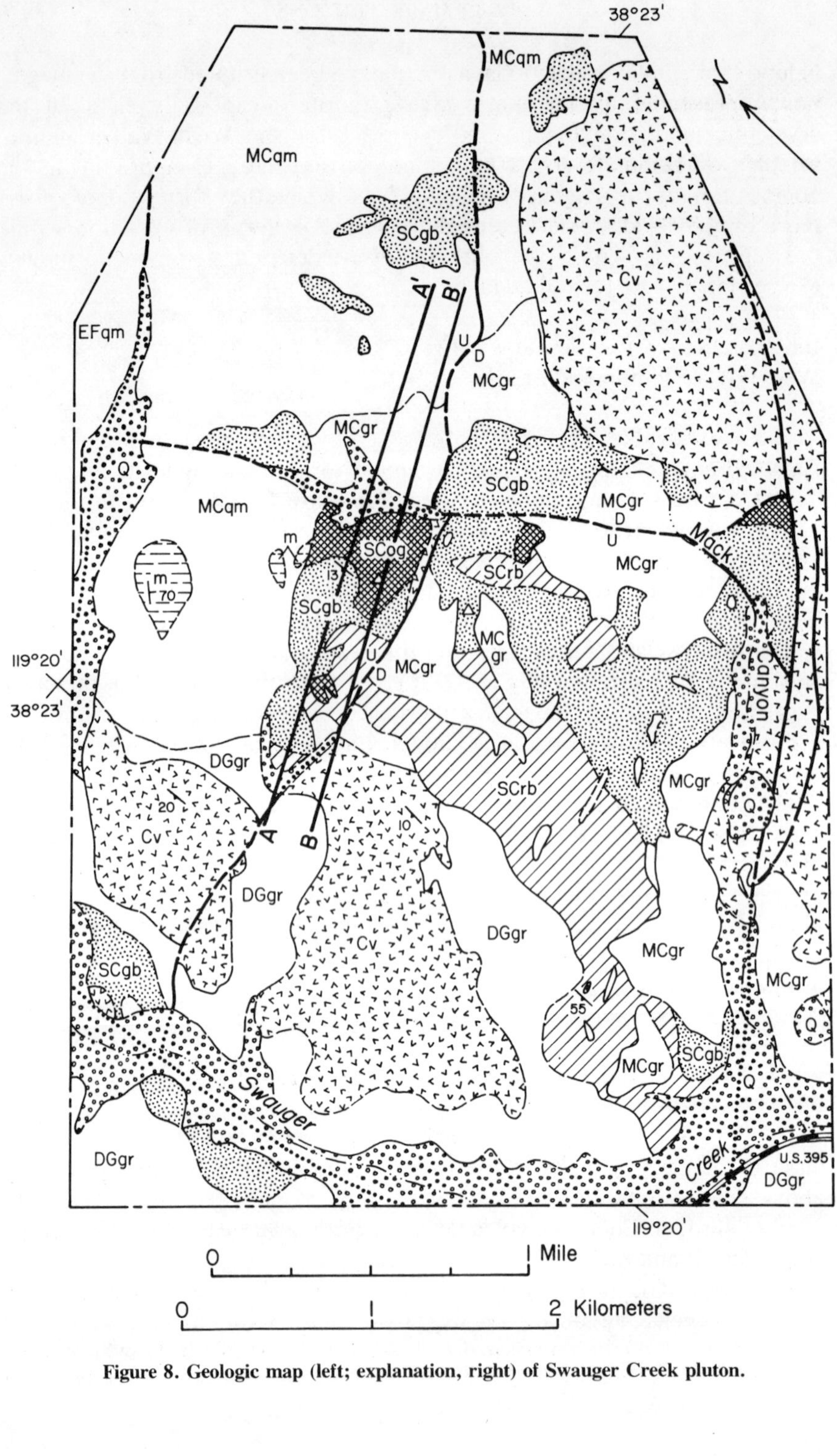

Figure 8. Geologic map (left; explanation, right) of Swauger Creek pluton.

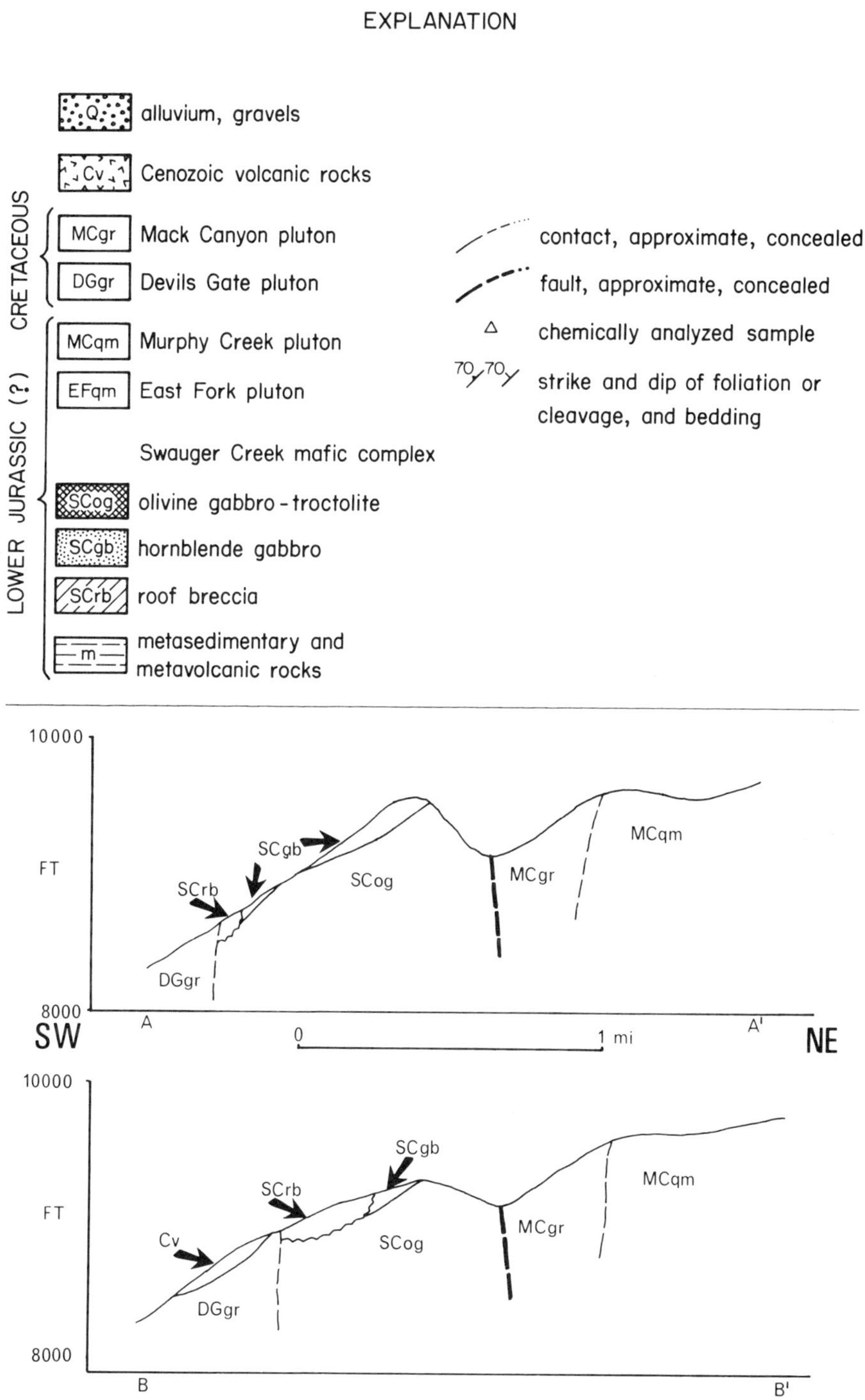

Figure 9. Cross sections through Swauger Creek pluton. Locations given in Figure 8.

is not exposed and whose layering now dips 10°W to 15°W but was probably originally horizontal before tilting. The apparent thickness, derived from cross sections (Fig. 9), is on the order of 200 m, but neither the roof nor the base of the body can be identified.

Vertical variation within the complex is great. The contacts shown in Figure 8 are completely gradational and can rarely be placed accurately owing to poor exposures. Melanocratic olivine-hornblende gabbro occurs at the lowest level of the complex. This unit grades upward through a strongly banded transition zone several metres thick into a unit consisting of hornblende gabbro that underlies the major exposed part of the complex. Upper parts of this latter unit become progressively more leucocratic and are quartz diorite and granodiorite. The hornblende gabbro unit grades in places into a complex, breccialike unit, with heterogeneous angular blocks of fine-grained gabbro, and hypabyssal and volcanic rocks of varied types embedded in a matrix of hornblende gabbro, quartz diorite, and granodiorite. The three general units are described in detail below.

Olivine Gabbro

Olivine gabbro crops out in four discontinuous masses, the largest of which underlies about 0.3 km^2 and is located about 2.5 km east of Swauger Creek and 3.2 km north of U.S. Highway 395 (Fig. 8). Two smaller bodies occur southeast of the large mass, adjacent to a northwest-trending normal fault, and a third body lies 0.5 km to the southwest. However, the 10°W to 15°W dip of the Swauger Creek complex indicates that olivine gabbro may underlie the entire exposed part of the complex, with a thin cap of hornblende gabbro overlying it (Fig. 9).

Deep green to brown in outcrop, the olivine gabbro is a mesh of poikilitic brown to green hornblende crystals as much as 3 cm across. These hornblende oikocrysts (Jackson, 1961) contain inclusions of rounded olivine (Fo_{74-86}) and subhedral bytownite that average 1.5 mm across, together with larger pleochroic pink bronzite anhedra which themselves enclose olivine and bytownite. Olivine enclosed by hornblende is typically jacketed by magnetite and is occasionally replaced by weblike clots of talc, iddingsite, and magnetite. Colorless clinopyroxene relicts, together with patches of anthophyllite, occur sparingly as inclusions in hornblende. Anthophyllite and pale bluish-green hornblende, often in zonal relationship to one another, also occur as randomly oriented fibrous sheaves. Green spinel (pleonaste), confined to areas in which the fibrous bluish-green hornblende occurs, locally forms both large anhedra (as much as 1 mm across) and symplectite-like intergrowths with the hornblende. The larger spinel anhedra frequently include opaque material around their edges. Tiny needles of deep blue riebeckite occur sparingly in some samples. Some rocks are rich in phyllosilicates that clearly replaced the amphiboles. Most frequently, coarse books of pale reddish-brown biotite and colorless to pale green chlorite are interleaved within host amphiboles. Talc has also been observed as coarse flakes in some rocks.

Modal analyses of these rocks are of little value as the original proportions of mineral constituents have been drastically modified by alterations and reactions. However, bytownite currently accounts for about 15 to 20 percent, and amphiboles, 50 to 60 percent of the rock by volume.

The chemical analysis of olivine gabbro reported in Table 3 is remarkably similar to Nockolds's average troctolite and may indicate that the olivine gabbro originated as an olivine-plagioclase rock.

The upper contact of the olivine gabbro (or troctolite) was mapped along a sharply banded zone consisting of alternating hornblende- and plagioclase-rich layers. The upper part of the olivine gabbro is characterized by an increase in modal bytownite and a decrease in modal olivine and orthopyroxene. Fibrous bluish-green hornblende and green spinel occur sporadically. Brown hornblende oikocrysts are slightly less abundant.

In the banded transitional zone, plagioclase is apparent in hand specimen. Hornblende oikocrysts, some with skeletal clinopyroxene inclusions, average 1 to 2 cm across. Clinopyroxene also forms discrete grains, some of which are twinned. Magnetite forms anhedra 3 to 4 mm across. Rocks texturally similar to these but with more abundant plagioclase form the bulk of the overlying hornblende gabbro.

Hornblende Gabbro and Related Rocks

Rocks mapped as hornblende gabbro form a rather heterogeneous, weakly foliated unit overlying the olivine gabbro. Modal compositions vary from hornblende gabbro to granodiorite, and great textural variation is present.

Most samples have a speckled appearance that is due to the development of green to brown hornblende oikocrysts which enclose subhedral to euhedral plagioclase crystals. These hornblende grains, as much as a centimetre across, have irregular outlines, owing to their interstitial nature. Some rocks with "jackstrawed" textures, characterized by random arrangement of green hornblende needles in hand specimen, grade into the more common speckled texture. Coarse hornblende pegmatites with crystals as much as 10 cm in length occur in irregular dikes, especially in the upper parts of the unit.

Plagioclase crystals are generally zoned. In the lowest parts of the unit, cores average about An_{77} and rims average An_{47}. The plagioclase becomes increasingly more sodic in the upper parts, and in granodiorite, cores average An_{39} and rims average An_{27}. Quartz and K-feldspar, absent in the lower parts where bytownite occurs, appear as conspicuous 1- to 2-cm-long oikocrysts enclosing andesine in quartz diorite and granodiorite of the upper levels. In these rocks, apatite and magnetite occasionally form large subhedral grains along with the plagioclase.

Alteration of the hornblende gabbro is not as pronounced as that of the olivine gabbro. Green biotite frequently replaces hornblende and often contains interleaved chlorite. Epidote, sometimes with green hornblende, forms coarse blebs in cores of plagioclase grains. In most rocks, ilmenite is rimmed with sphene.

Evidence of Crystal Settling

Two lines of evidence suggest that the olivine gabbro and overlying hornblende gabbro and related rocks formed by gravitative settling of early crystallized phases from a basaltic magma. First is the vertical mineral variation in the units summarized above, with ultramafic rocks at the base grading upward into progressively more silicic rocks. Second is the textural evidence of cumulus and intercumulus phases. Many of the textures of these rocks are similar to those illustrated for the Skaergaard Complex by Wager and Brown (1967).

Plagioclase is a cumulus phase or primocryst (Wager and Brown, 1967) throughout the body and is associated with olivine and clinopyroxene in the olivine gabbro. Hornblende gabbro contains rare cumulus apatite, and hornblende together with the plagioclase forms primocrysts near the exposed top of the intrusion. Whereas hornblende is the major intercumulus phase throughout the hornblende gabbro, K-feldspar and quartz make an appearance as intercumulus oikocrysts near the top, where hornblende is a cumulus phase. At any one place in the body, the same mineral was never observed to be both an intercumulus and a cumulus phase.

The order of appearance and disappearance of cumulus phases and the nature of the intercumulus oikocrysts observed in the Swauger Creek complex are consistent with fractionation and crystal settling in a basaltic magma. The liquid line of descent, deduced from mineral parageneses at various levels within the complex, was as follows: gabbro → diorite → quartz diorite → granodiorite. This suggests that a common basalt magma, trapped in a small chamber, fractionated to produce a final product texturally and chemically similar to many of the small mafic bodies of the eastern Sierra Nevada.

Roof Breccia

Widespread younger intrusions have obliterated the original walls and roof of the Swauger Creek complex (Fig. 8). However, proximity to the original roof of the complex is suggested by large zones containing angular blocks of diverse lithologies embedded in intrusive rock. These occur in isolated masses on Swauger Creek, along the west edge of the main mass, and as isolated pods in hornblende gabbro. This extremely heterogeneous rock is termed a roof breccia and is composed of blocks of fine-grained gabbro, diabase, and porphyritic metavolcanic rocks interlaced with dikelike apophyses of granodiorite and jackstrawed gabbro. The small pods of breccia isolated within hornblende gabbro are similar to the main mass of breccia, except that the interstices between the angular fragments are occupied by more gabbroic rock. Lithologies of angular fragments are greatly variable, even within individual outcrops. However, most fragments appear to be porphyritic andesitic or basaltic lava with less abundant fine-grained gabbro and diorite. Some blocks of metasedimentary rock occur in roof breccia north of the east fork of Swauger Creek (Fig. 2).

Interpretation

The Swauger Creek mafic complex apparently formed from a basaltic magma that fractionated in a high-level magma chamber and settled out olivine and calcic plagioclase crystals. The last liquid to crystallize formed small amounts of quartz diorite or granodiorite near the top of the magma chamber. Parts of the volcanic roof of the complex collapsed into the still partly liquid mass and were intermixed to varying degrees. In some areas, blocks and fragments are attenuated; this suggests that temperatures were high enough for some of the material to deform plastically. Volcanic roof rocks beneath which the sill was emplaced probably were of Early Jurassic age, although none are preserved in the immediate vicinity of the intrusion. Fine-grained gabbroic rocks in the roof breccia may represent parts of the magma that were chilled near the roof of the magma chamber.

Geochronology

R. L. Armstrong performed K-Ar analyses of 12 biotite and hornblende separates from 10 samples collected from five plutons in the area studied. These data, together with calculated ages, are listed in Table 5. With the exception of one anomalously discordant biotite-hornblende pair (with biotite age older than hornblende—sample 104), the calculated ages fall in the range of 76 to 93 m.y. This fact is remarkable, considering that the radiometric data do not allow any clear distinction between the apparent ages of the oldest (Swauger Creek) and the youngest (Devils Gate and Mack Canyon) plutons, which geologic criteria indicate are Early Jurassic and Late Cretaceous, respectively. Furthermore, this spread of ages closely parallels the interval 77 to 90 m.y., termed by Evernden and Kistler (1970) the Cathedral Range intrusive epoch.

Figure 10 illustrates the geometric relationships between the plutons studied here and the larger plutons of the Cathedral Peak type which have been extensively dated by Evernden and Kistler. Although a biotite K-Ar age of 119 m.y. was reported by Evernden and Kistler (1970) for a sample collected from the Mill Creek pluton (their sample no. 1697), outcrops of this body can be followed without break into exposures of the lithologically identical Sonora pluton, which yielded 14 K-Ar ages in the range of 79.5 to 85.6 m.y. (Evernden and Kistler, 1970). Therefore, the lithologically identical rocks of the Mill Creek and Rock Creek plutons probably represent offshoots of the large Upper Cretaceous intrusions of Cathedral Peak–type quartz monzonite.

The apparent equivalence of the K-Ar ages of samples from the China Garden, Desert Creek, and Swauger Creek plutons with the Cathedral Range intrusive epoch confirms the petrographic observations that biotite and hornblende in these rocks are late, postmagmatic, and probably metamorphic. These young ages clearly reflect the later heating and cooling

→

Figure 10. Map showing proximity of older plutons of the study area to plutons of the Cathedral Range intrusive epoch. Data from Evernden and Kistler (1970), Clark (1968), Stanford Geological Survey (1961, 1963, 1964, 1966, 1968, unpub. maps and repts. of summer field classes, W. R. Dickinson in charge), Koenig (1963), and this study.

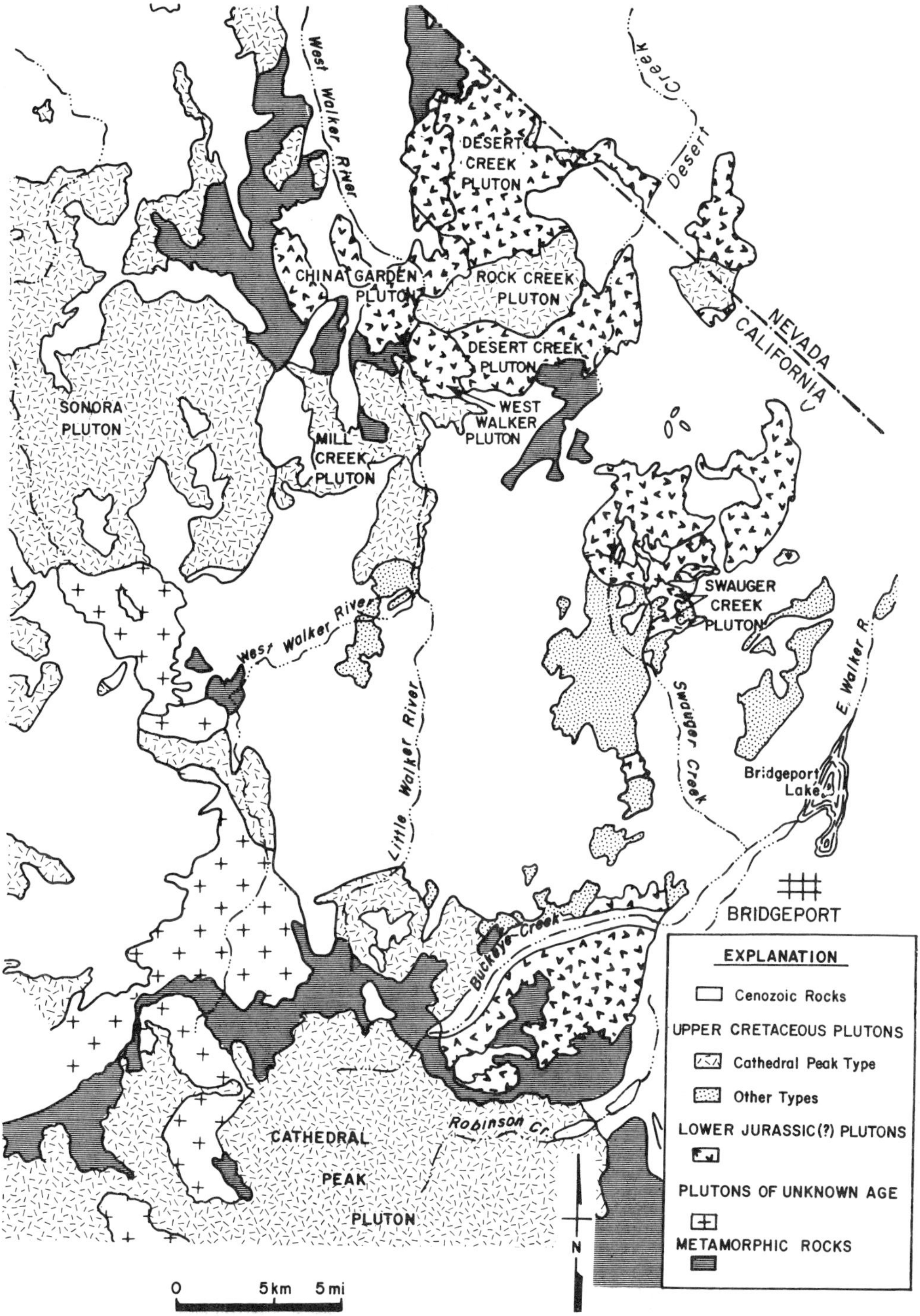

West Walker River
DESERT CREEK PLUTON
Desert Creek
NEVADA
CALIFORNIA
CHINA GARDEN PLUTON
ROCK CREEK PLUTON
DESERT CREEK PLUTON
WEST WALKER PLUTON
SONORA PLUTON
MILL CREEK PLUTON
West Walker River
SWAUGER CREEK PLUTON
E. Walker R.
Little Walker River
Swauger Creek
Bridgeport Lake
BRIDGEPORT
Buckeye Creek
Robinson Cr.
CATHEDRAL PEAK PLUTON
N
0
5 km
5 mi
EXPLANATION
Cenozoic Rocks
UPPER CRETACEOUS PLUTONS
Cathedral Peak Type
Other Types
LOWER JURASSIC(?) PLUTONS
PLUTONS OF UNKNOWN AGE
METAMORPHIC ROCKS

TABLE 5. K-Ar ANALYSES

Pluton and sample no.	Mineral	K 1 (%)	K 2 (%)	Ar* [× 10^{-6} cm³ (%)]	Age (m.y.)
Swauger Creek					
SC4	Biotite	7.05	7.09	22.77 (8)	79.0 ± 1.6
	Hornblende	1.47	1.47	5.01 (39)	83.5 ± 1.7
SC32A	Hornblende	0.412	0.407	1.56 (73)	93.0 ± 3.4
SC41	Hornblende	0.633	0.635	2.37 (63)	91.4 ± 2.5
China Garden					
218	Biotite	5.28	5.37	17.44 (27)	80.3 ± 1.6
176	Biotite	7.28	7.24	23.78 (31)	80.3 ± 1.6
104	Biotite	5.83	5.63	17.70 (35)	75.8 ± 1.5
	Hornblende	0.195	0.209	0.490 (90)	59.8 ± 5.8
				0.505 (60)	61.6 ± 2.0
Desert Creek					
241	Hornblende	0.535	0.541	1.78 (81)	81.3 ± 4.3
107	Hornblende	0.451	0.471	1.63 (59)	86.3 ± 2.1
Devils Gate					
DG	Biotite	4.02	3.96	13.58 (43)	83.4 ± 1.7
Mack Canyon					
SC10	Biotite	7.01	6.89	22.34 (22)	78.8 ± 1.6

Note: Analyses provided by R. L. Armstrong. Sample locations: SC4—lat 38°21′N, long 119°20′W. SC32A—lat 38°22′N, long 119°19′W. SC41—lat 38°22′N, long 119°19′W. 218—lat 38°30.5′N, long 119°29.3′W. 176—lat 38°31.3′N, long 119°25′W. 104—lat 38°29′N, long 119°19′W. 241—lat 38°28′N, long 119°24.5′W. 107—lat 38°32.5′N, long 119°23′W. DG—lat 38°21′N, long 119°22.5′W. SC10—lat 38°21′N, long 119°19.5′W.

*Determined values of Ar given at standard temperature and pressure. The number in parentheses is the percent air correction.

of these older plutons that correspond to the intrusion and subsequent cooling of the Cathedral Peak-type plutons. A similar example of K-Ar ages of an old pluton reset by younger intrusions can be found in Table 5 of Evernden and Kistler (1970). Of eight biotite and hornblende K-Ar ages from the Upper Triassic Mono Dome granodiorite, seven (five biotites and two hornblendes) fall in the range of 82 to 99 m.y., whereas one (hornblende) yielded a Late Triassic age of 206 m.y. Many other examples of reset ages of old plutons and discordant mineral pairs are apparent in Plate 2 of Evernden and Kistler's paper and were cited by them as evidence that the older plutonic rocks were reheated by the emplacement of the younger plutons.

The K-Ar ages reported here unfortunately do not afford a basis for assigning absolute ages to the older plutons in the Walker drainage. Instead, the data provide further evidence that the older plutons were reheated and suggest that many of the characteristics of alteration and metamorphism of these plutons described above resulted from intrusion of the Upper Cretaceous plutons adjacent to and among them (Fig. 10).

Modal, chemical, and normative data (discussed below), together with the probable genetic relationships between the older plutons and Lower Jurassic metavolvanic rocks, provide more substantial evidence that the older, metamorphosed plutons in the Walker drainage area may be counterparts of Lower Jurassic plutons in the White and Inyo Mountains to the south.

Regional Correlations

Mafic intrusions, Upper Triassic to Middle Jurassic felsic plutons, and remnants of Upper Triassic to Lower Jurassic volcanic and sedimentary rocks are distributed in subparallel, northwest-trending belts in eastern California (Fig. 11). The three belts lie mainly east of, but partly within the east edge of, the Cretaceous part of the Sierra Nevada batholith. The petrology, chemistry, and geometry of these rocks argue that the belts of coeval intrusive and extrusive rocks are different expressions of a marginal magmatic arc of Late Triassic to Middle Jurassic age.

MAFIC INTRUSIONS

The West Walker and Swauger Creek plutons (locs. A and B in Fig. 11) are the only mafic intrusions that have been studied in detail in the eastern Sierra Nevada. However, other mafic bodies that have been described are lithologically and structurally similar to these two bodies (Moore, 1963; Sherlock and Hamilton, 1958; Rinehart and Ross, 1964; Noble, 1962). The general distribution of these small mafic plutons is roughly coextensive with the distribution of remnants of metamorphosed sedimentary and volcanic rocks, and according to Bateman and others (1963), the mafic intrusions generally predate and are metamorphosed by adjacent felsic bodies.

Except for the Swauger Creek pluton, none of the mafic bodies has been dated radiometrically, and they may range considerably in age. In the Walker drainage area, however, the mafic plutons are older than the probable Lower Jurassic felsic plutons; many, if not most, of the mafic plutons of the eastern Sierra Nevada may be Triassic or Jurassic as well.

FELSIC PLUTONS

Felsic plutons similar in chemistry, petrology, and probably age to those described in this paper have been recognized over much of eastern California and western Nevada (Fig. 11). Sylvester and others (1969) described hornblende and augite monzonites in the Inyo Mountains. Early Jurassic ages were reported for these rocks and associated quartz monzonites by McKee and Nash (1967), and younger quartz monzonites yielded Cretaceous ages. Dunne (1971) described a pyroxene-bearing seriate monzonite-quartz monzonite pluton (the Pat Keyes pluton) in the southern Inyo Mountains, which yielded a Rb-Sr isochron age of 183 m.y. Dunne (1971) and Ross (1965) considered this pluton to be a part of the larger Hunter Mountain batholith to the southeast, also of Early Jurassic age (Ross, 1965). Mapped by McAllister (1956), the batholith is dominantly quartz monzonite, but contains local pyroxene-bearing monzonitic and syenitic variants. Hall and MacKevett (1962) gave radiometric ages of 180 to 210 m.y. for rocks that occur in the Argus Range and that are petrographically similar to those of the Hunter Mountain batholith, 80 km to the northwest (Fig. 11). Evernden and Kistler (1970) reported Early to Middle Jurassic ages for seriate-textured pyroxene-bearing granodiorite described by Bateman (1961) east of the Mount Goddard pendant (Fig. 11). To the north, granodiorite and quartz monzonite near Mono Lake yielded maximum K-Ar ages of 210 m.y. and a Rb-Sr isochron age of 206 m.y. (Evernden and Kistler, 1970). Castor and McKee (1975) recently reported an Early Jurassic K-Ar hornblende age of 180 m.y. from a pluton in the Pine Nut Range (Fig. 11).

The terrane partly underlain by plutons that range in age from Late Triassic to Early and Middle Jurassic thus extends from the Pine Nut Range more than 400 km southward to the Garlock Fault (Fig. 11). Additional Triassic and Jurassic plutons occur in western Nevada, but their distribution has not been completely determined (Smith and others, 1971; Silberman and McKee, 1971; Evernden and Kistler, 1970). Evernden and Kistler (1970) reported K-Ar ages of 151 to 162 m.y. on one pluton in the western Sierra Nevada (loc. C in Fig. 11), and this has led Kistler and others (1971) and Kistler (1974) to regard that pluton as part of a more northwesterly Jurassic plutonic trend. The apparent K-Ar ages make this pluton early Late Jurassic rather than Early or Middle Jurassic; in any case, the discordance of biotite and hornblende ages for this pluton (Evernden and Kistler, 1970) and the limited data do not at present allow a meaningful interpretation of the age or significance of this pluton (Schweickert and Cowan, 1975).

COEVAL METAVOLCANIC AND METASEDIMENTARY ROCKS

Most remnants of metamorphosed sedimentary and volcanic rocks in eastern California and western Nevada lie slightly west of but parallel to the belt of older felsic plutons (Fig. 11). Permian-Triassic and Lower Triassic volcanic rocks are not considered here, because their tectonic

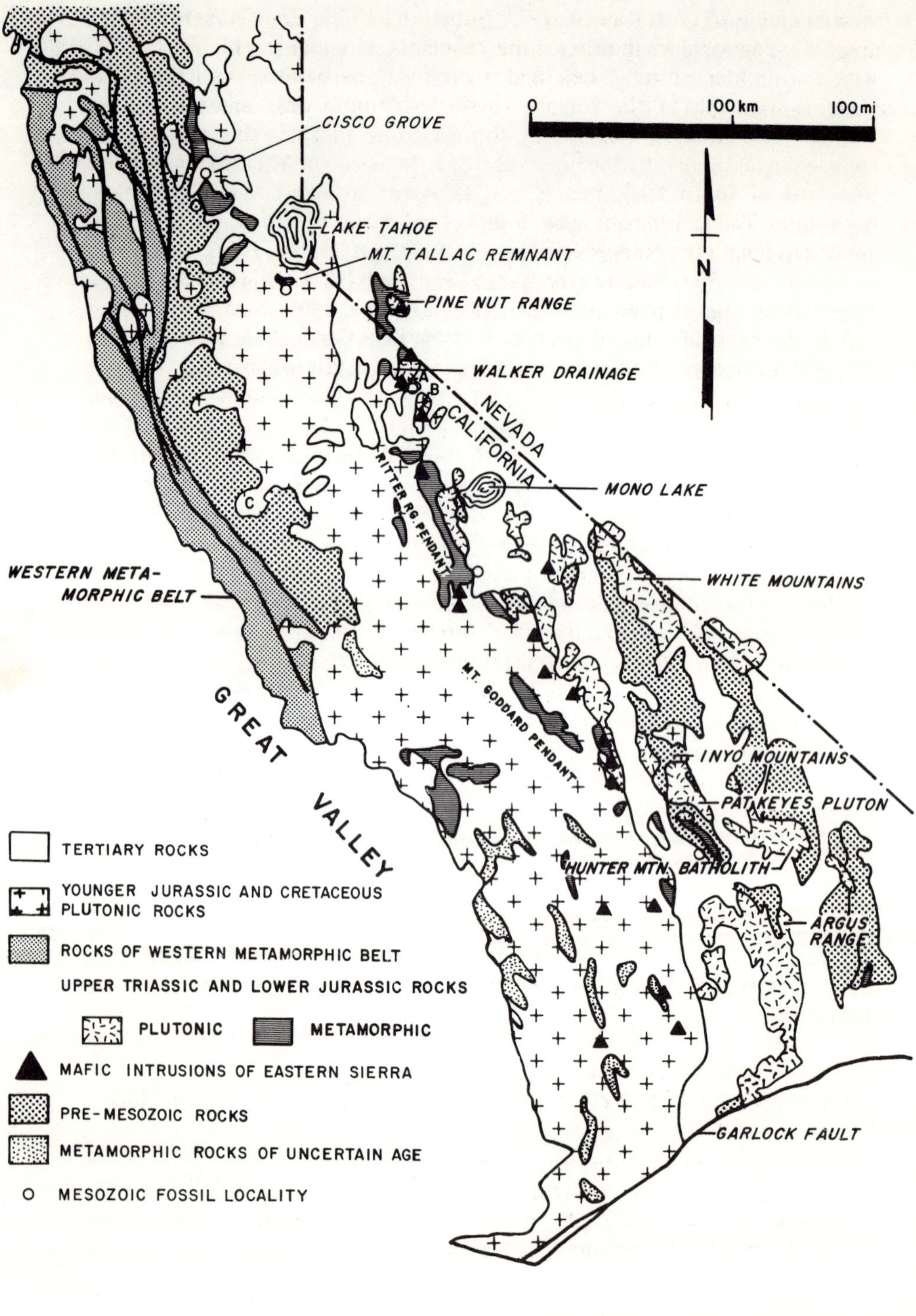
0
100 km
100 mi
N
CISCO GROVE
LAKE TAHOE
MT. TALLAC REMNANT
PINE NUT RANGE
WALKER DRAINAGE
A
B
C
NEVADA
CALIFORNIA
MONO LAKE
RITTER RG. PENDANT
WHITE MOUNTAINS
WESTERN META-
MORPHIC BELT
MT. GODDARD PENDANT
GREAT
VALLEY
INYO MOUNTAINS
PAT KEYES PLUTON
HUNTER MTN. BATHOLITH
ARGUS
RANGE
GARLOCK FAULT
TERTIARY ROCKS
YOUNGER JURASSIC AND CRETACEOUS
PLUTONIC ROCKS
ROCKS OF WESTERN METAMORPHIC BELT
UPPER TRIASSIC AND LOWER JURASSIC ROCKS
PLUTONIC
METAMORPHIC
MAFIC INTRUSIONS OF EASTERN SIERRA
PRE-MESOZOIC ROCKS
METAMORPHIC ROCKS OF UNCERTAIN AGE
MESOZOIC FOSSIL LOCALITY

significance is unclear (Speed, 1973; Silberling, 1973). The Walker drainage area is exceptional in that here, the remnants of metamorphic rock occur *within* the older plutonic belt and are closely associated with the older plutons. Throughout the region, fossils are found only sparsely in the metamorphic rocks. Nearly all of the Mesozoic rocks in this narrow belt have been assigned to the interval Late Triassic to Middle Jurassic on the basis of fossil finds near Cisco Grove (Clark and others, 1962), in the Mount Tallac remnant near Fallen Leaf Lake (Loomis, 1961), in the southern Pine Nut Range (Noble, 1962; D. C. Noble 1969, 1970, oral communs.), in the Sweetwater Range near Lobdell Lake (Halsey, 1953), in the Ritter Range pendant (Rinehart and others, 1959), and at the south end of the Inyo Mountains near Lone Pine (Merriam, 1963; Knopf, 1918) (Fig. 11). Lithologically similar metavolcanic suites of probable Late Triassic age also occur in the Mojave Desert (Abbott, 1971; Rogers and others, 1974).

Published lithologic descriptions of the metamorphic rocks within this belt have much in common. In general, intermediate to silicic pyroclastic rocks with subaerial welded tuffs, interlayered with less abundant mafic lava flows, are the predominant lithologic types, although this varies from pendant to pendant. Nearly all pendants and roof remnants also contain epiclastic rocks in variable amounts, sporadically distributed throughout the Mesozoic section (Bateman and others, 1963). Such epiclastic rocks always show a strong volcanic contribution and are best termed volcaniclastic rocks.

The metamorphic section in the Walker drainage area contains a large proportion of epiclastic conglomerate, sandstone, and siltstone in its lower part. Very rough estimates of thickness are shown in Figure 3. Associated with the older plutons, however, are thick accumulations of felsic to intermediate pyroclastic rocks and lava flows.

OLDER AND YOUNGER BATHOLITH BELTS

Some geologists, noting that Upper Triassic-Middle Jurassic plutons occur both in the Inyo and White Mountains and in the eastern Sierra Nevada, have argued that batholiths in these areas are comagmatic (McKee and Nash, 1967; Ross, 1965). In contrast, Dunne (1971) noted the low quartz content and great lithologic variation in the Pat Keyes pluton and Hunter

Figure 11. Lower Jurassic and Upper Triassic rocks of eastern California. Areas shown as Lower Jurassic and Upper Triassic metamorphic rocks also contain some Paleozoic and Permian-Triassic rocks which are not differentiated.

Mountain batholith and concluded that although some plutons in the Sierra Nevada and Inyo Mountains may be coeval, they are not comagmatic. Sylvester and others (1969), noting the presence of monzonitic rocks in the Inyo batholith and their absence in the Sierra Nevada batholith, suggested that not all rocks in the Inyo batholith are comagmatic with those of the Sierra Nevada batholith.

Part of the confusion here results from terminology. Rocks in the Inyo and White Mountains are often referred to the "Inyo batholith," whereas those in the Sierra Nevada are all included in the "Sierra Nevada batholith." Unlike the "Sierra Nevada batholith," the "Inyo batholith" actually consists of many batholith-sized intrusive masses that are physically separated by broad expanses of Paleozoic and Mesozoic rocks at the present level of exposure (Fig. 11). Nevertheless, both ranges contain plutonic rocks intruded in the two distinctly different time intervals, Late Cretaceous and Late Triassic–Middle Jurassic. Lower Cretaceous and Upper Jurassic plutons of intermediate age mainly occur west of the crest of the Sierra Nevada.

Plutons in the two adjacent mountain ranges are more logically grouped according to their radiometrically determined ages. Thus, Triassic and Jurassic plutons should be grouped separately from Upper Cretaceous plutons, regardless of where they occur. Such age groupings enable the consideration of petrologic features of plutons that were emplaced and subsequently solidified during the same general magmatic pulse (Evernden and Kistler, 1970). This format was adopted by Bateman and Dodge (1970) for their discussion of chemical patterns of granitic rocks of the Sierra Nevada region.

COMPARISONS BETWEEN OLDER AND YOUNGER PLUTONIC ROCKS

Modes

When Upper Triassic to Middle Jurassic plutons that occur in the two mountain ranges are considered together and compared with Upper Cretaceous plutons of both ranges, several petrologic distinctions can be made. Modes reported in this paper (Tables 1 and 2) and modes reported by Bateman (1961), Dunne (1971), McAllister (1956), Hall and MacKevett (1962), Moore (1963), and Ross (1965) all indicate appreciably less modal quartz (averaging 15 to 25 percent by volume) in the older rocks than in Upper Cretaceous plutons (commonly 25 to 35 percent modal quartz by volume) with which the older plutons are associated.

Figure 12 shows this comparison. In addition, many of the demonstrably older plutons contain clinopyroxene as an important mafic constituent, whereas most younger plutons contain clinopyroxene only as rare cores in hornblende (Bateman and others, 1963).

Chemistry

Evernden and Kistler (1970) termed their proposed Late Triassic and Early to Middle Jurassic intrusive epochs the Lee Vining and Inyo Mountains intrusive epochs, respectively. Bateman and Dodge (1970) have shown that chemical variation trends for plutonic rocks of these epochs are somewhat divergent from those of Cretaceous and Upper Jurassic intrusions.

Chemical analyses of quartz monzonite of the China Garden and Desert Creek plutons and mafic quartz diorite of the West Walker pluton (Table 3) are plotted together with the data of Bateman and Dodge in Figure 13. The chemically heterogeneous Upper Triassic rocks are excluded. In

Figure 12. Average modes of selected plutons (Q, quartz; K, K-feldspar; P, plagioclase). Data from Bateman (1961), Dunne (1971), Hall and MacKevett (1962), Kistler (1966b), McAllister (1956), Moore (1963), Noble (1962), Rinehart and Ross (1964), Ross (1965, 1969), and Smith and others (1971).

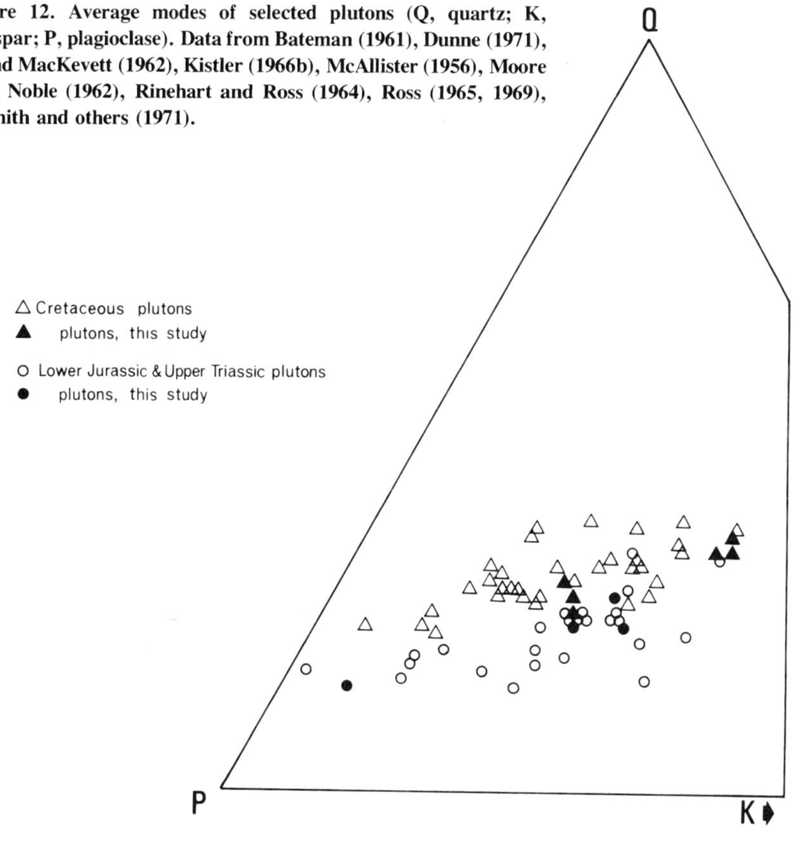

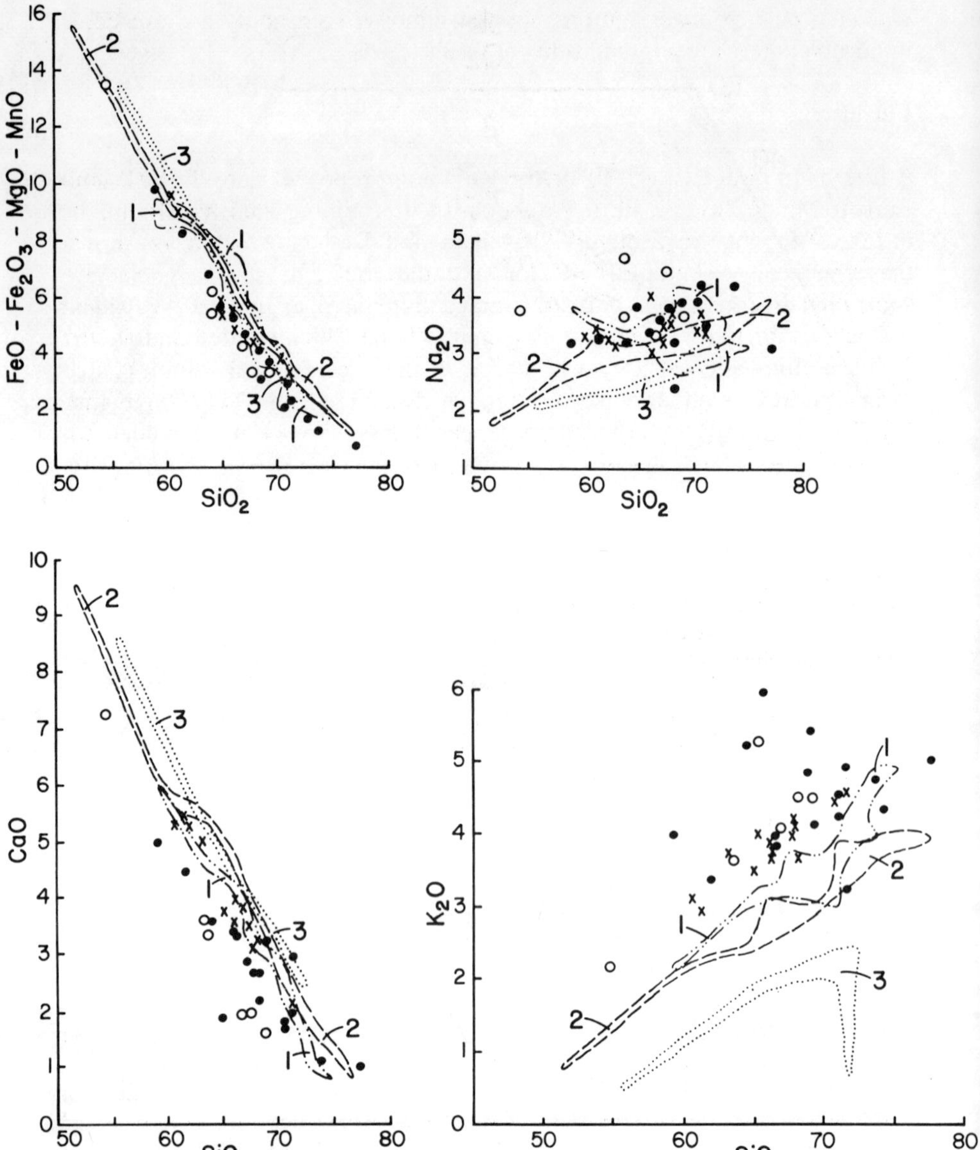

Figure 13. Chemical comparisons between older plutons in the Walker drainage area and other Sierra Nevada plutons (modified from Bateman and Dodge, 1970, Figs. 5, 7). Open circles, analyses of plutons in the study area; dots, Lower and Middle Jurassic White Mountains rocks; x's, Lower and Middle Jurassic Palisade Crest sequence; area 1, Upper Cretaceous sequences; area 2, Lower Cretaceous sequences; area 3, Upper Jurassic sequences. See text for discussion.

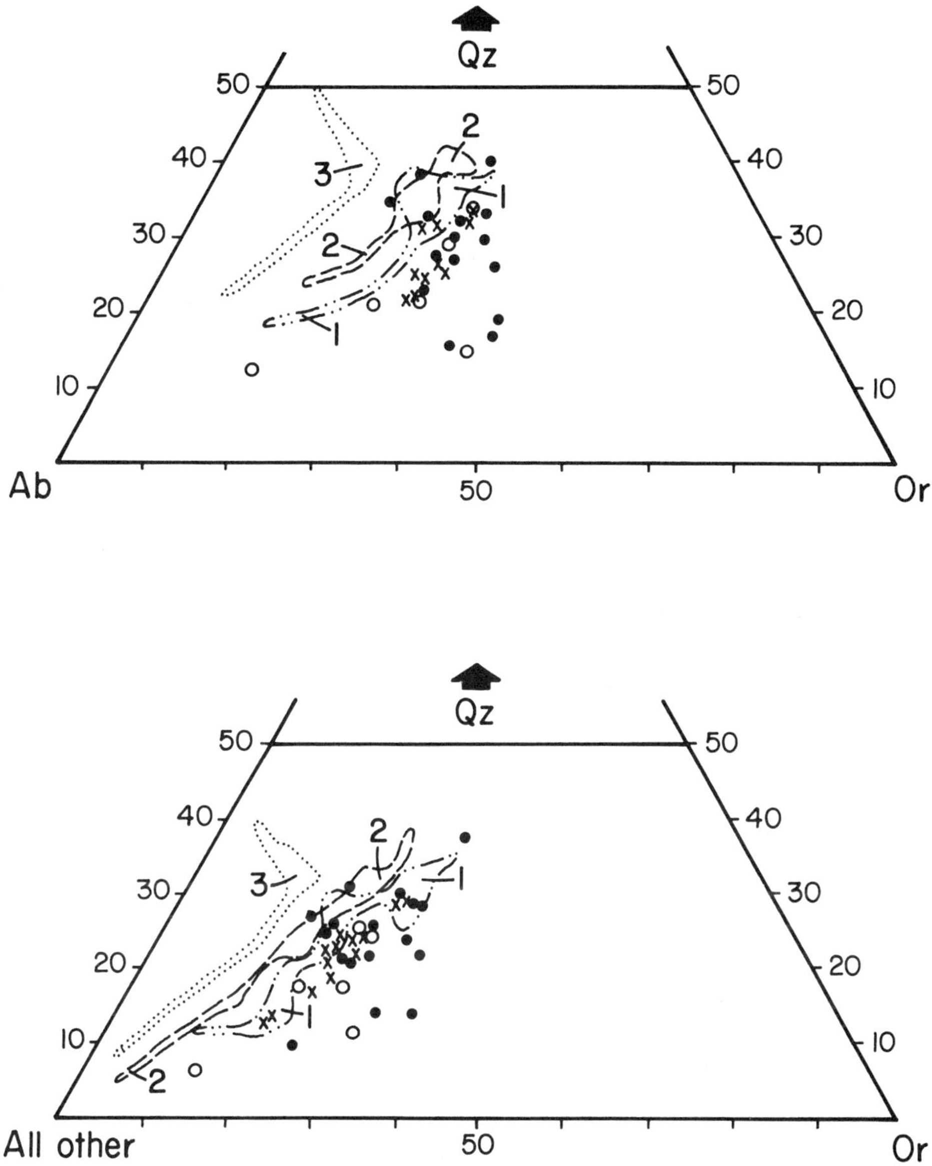

Figure 14. Normative comparisons between older plutons in the Walker drainage area and other Sierra Nevada plutons (modified from Bateman and Dodge, 1970, Fig. 8). Symbols as in Figure 13. Ab, albite; Qz, quartz; Or, orthoclase. See text for discussion.

general, the older sequences contain more K_2O and less CaO at a given SiO_2 percentage than do the Cretaceous and Upper Jurassic sequences.

On all four plots in Figure 13, the analyzed rocks of the Walker drainage area closely follow the chemical patterns of the Lower Jurassic sequences and are distinct from those of the younger sequences. This is especially true for CaO versus SiO_2 and K_2O versus SiO_2. In addition, the normative plots of Figure 14 show that the rocks of the Walker drainage area are low in normative quartz and high in normative orthoclase, exactly like the Lower Jurassic sequences. These chemical similarities, together with the modal data, support the inference that the China Garden and Desert Creek plutons are part of the belt of Lower Jurassic intrusive rocks.

Contrasting K_2O-SiO_2 Variations

As mentioned above, these older intrusive sequences are more potassic than younger sequences. In addition, K_2O-SiO_2 plots for most of the older rocks exhibit much greater scatter than those of the younger sequences (Bateman and Dodge, 1970, Figs. 5 and 7). Mafic rocks in the eastern Sierra Nevada may also contain more K_2O than their younger counterparts in the western Sierra Nevada, but analytical data are not abundant.

Noting this great variability of K_2O/SiO_2 ratios of the Lower Jurassic White Mountains rocks and the Upper Triassic Scheelite sequence, Bateman and Dodge suggested that late magmatic or postmagmatic vapor-phase transport of K may have occurred in these older K-rich sequences. As discussed earlier, this mechanism may be operative in shallow-level intrusions and may be independent evidence that many of the plutons of the older sequences, like those described in this paper, are high-level or epizonal intrusions.

Dickinson (1970), using chemical data of Bateman and Dodge (1970), argued that K_2O variations across the central Sierra Nevada batholith implied the former existence of an eastward-dipping Mesozoic Benioff zone beneath the continental margin. Rocks of the older sequences with great variability in K_2O were omitted from Dickinson's plot, but data from the Lower Jurassic Palisade Crest sequence of Bateman and Dodge, which shows less scatter, were used.

The fact that the K-rich older sequences east of the Sierra Nevada crest appear to continue the eastward increase of K_2O in granitic rocks across the batholith—even though they paradoxically exhibit the oldest ages and are chemically most unlike the next younger Upper Jurassic intrusive sequence in the western Sierra Nevada foothills—probably is coincidental and may instead be related mainly to the fact that the oldest intrusions are presently exposed at higher levels than are younger parts of the Sierra Nevada batholith (discussed below).

Levels of Emplacement

In addition to the plutons discussed in this paper, other plutons in the eastern belt show evidence of shallow emplacement. Dunne (1971) observed that the Pat Keyes pluton, which yielded a 183-m.y. Rb-Sr isochron age, intrudes a section of Triassic rocks 1.3 km thick; he argued that the pluton must therefore have solidified under a rather thin cover. Border facies of the Pat Keyes pluton may be analogous to those of the China Garden and Desert Creek plutons. Most of the Triassic and Jurassic plutons that lie east of the Sierra Nevada crest form discrete intrusive masses that are completely encased in metamorphosed sedimentary and volcanic rocks (Fig. 11). Ross (1965) suggested in his study of the Hunter Mountain batholith that this fact may indicate that most of these older plutons are incompletely unroofed.

In contrast, plutons in the interior of the Sierra Nevada batholith form a crude mosaic with little intervening metamorphic roof and wall rocks exposed at the present level of erosion. Only a few sparse roof remnants of older metavolcanic and metasedimentary rocks are found among the younger plutons. Nearly all estimates of the depth of emplacement for the Cretaceous plutons have exceeded 5 km (Loomis, 1966; Evernden and Kistler, 1970), and some estimates have been greater than 10 km (Bateman and others, 1963).

Volcanic rocks of Cretaceous age, presumed eruptive equivalents of the extensive Upper Cretaceous batholiths, have not been reported in the Sierra Nevada region[1] and instead are probably represented by thick successions of Cretaceous volcanogenic sedimentary rocks of the Great Valley sequence which flank the batholith on the west (Ojakangas, 1968; Dickinson and Rich, 1972). This too suggests that the Upper Cretaceous batholiths are more deeply eroded than adjacent Upper Triassic to Middle Jurassic plutons.

TECTONIC SETTING

The parallel arrangement of the linear belts of coeval Upper Triassic to Middle Jurassic intrusive and extrusive rocks (Fig. 11) suggests that an Andean-type volcano-plutonic magmatic arc existed along the continental margin in eastern California and western Nevada during that period. This suggestion has been made previously by Stanley and others (1971), who

[1]R. W. Kistler (1975, written commun.) stated that Lower Cretaceous metavolcanic rocks occur in the Ritter Range, Mount Goddard, Strawberry, and Merced Peak pendants, as determined by Rb-Sr age dating.

termed it an "island arc," and by Burchfiel and Davis, 1972, who called it an "Andean arc." The preponderance of intermediate to silicic pyroclastic rocks, with subordinate amounts of andesite and basalt, the occurrence of large quartz monzonite and granodiorite batholiths, and the underlying Precambrian and Paleozoic basement clearly are more consistent with an Andean analogue, for as Hamilton (1969a) has shown, the central Andes are characterized by a broad plateau of silicic to intermediate ignimbrites mounted by dacitic stratovolcanoes, with a large upper Cenozoic batholith inferred at shallow depth beneath. Hamilton (1969a) offered the volcanic central Andes as an actualistic model for the Cretaceous batholiths of western North America, and this model seems appropriate as well for the older arc terrane discussed here.

However, as Burchfiel and Davis (1972) pointed out, this early Mesozoic arc is situated at a large angle to trends of isopachs and lithofacies and instead parallels the truncated edge of the Paleozoic Cordilleran orogen. Thus, the arc was rooted in sialic crust of Precambrian crystalline and Paleozoic shelf rocks to the southeast, Paleozoic miogeoclinal rocks in the White Mountains, Inyo Mountains, and Argus Ranges and presumably developed on more "oceanic" Paleozoic eugeosynclinal rocks in areas farther northwest. This longitudinal transition in character of underlying crustal rocks may be reflected in the nature of metavolcanic rocks of the magmatic arc, in that the more northerly occurrences of Upper Triassic to Middle Jurassic rocks (at Taylorsville, Sierra City, Cisco Grove, and the American River; Fig. 15) contain little evidence of subaerial silicic volcanism as in areas to the south, but instead are characterized by thick successions of marine lava flows, tuff, and breccia of dominantly andesitic composition (Diller, 1908; McMath, 1966; Crickmay, 1933; Clark and others, 1962; Stuart-Alexander, 1966; Loomis, 1961; Schweickert, unpub. data). In other words, the Triassic to Jurassic magmatic arc identified herein may show a transition from Andean (continental) character in its southern part to Indonesian or Japanese (oceanic) character in its northern part, in a manner entirely analogous to the Aleutian-Alaska Peninsula arc terrane that is constructed partly on oceanic and partly on continental crust (Forbes and others, 1969). Sr-isotope and trace-element data on plutonic rocks reported by Kistler and Peterman (1973) also show longitudinal variations, with more primitive initial Sr values occurring to the northwest, where plutons invaded "oceanic" Paleozoic rocks.

Data from the California Coast Ranges indicate that underthrusting and metamorphism of Franciscan mélanges began during Late Jurassic time (Suppe, 1969, 1973; Lanphere, 1971). In fact, sedimentary rocks older than Late Jurassic have not been recognized west of the Great Valley. It is likely, therefore, that tectonic deformation of the Franciscan Complex during subduction, as envisioned by many geologists (Ernst, 1970; Suppe,

1969, 1973; Cowan, 1974), is related only to the Late Jurassic to Late Cretaceous Sierra Nevada plutonism. Additionally, it is likely that the pre-Late Jurassic magmatism of eastern California and western Nevada discussed in this paper must be related to subduction of oceanic crust and lithosphere elsewhere before the subduction of the Franciscan Complex occurred.

The belt of steeply east-dipping fault zones in the foothills of the western Sierra Nevada (Clark, 1964; Figs. 11 and 15) may represent a subduction zone that was linked to the development of the Late Triassic to Middle Jurassic magmatic arc and that thus predated the subduction zone involving the Franciscan Complex (Schweickert and Cowan, 1975). These fault zones, like the Coast Range thrust of Bailey and others (1970), are generally

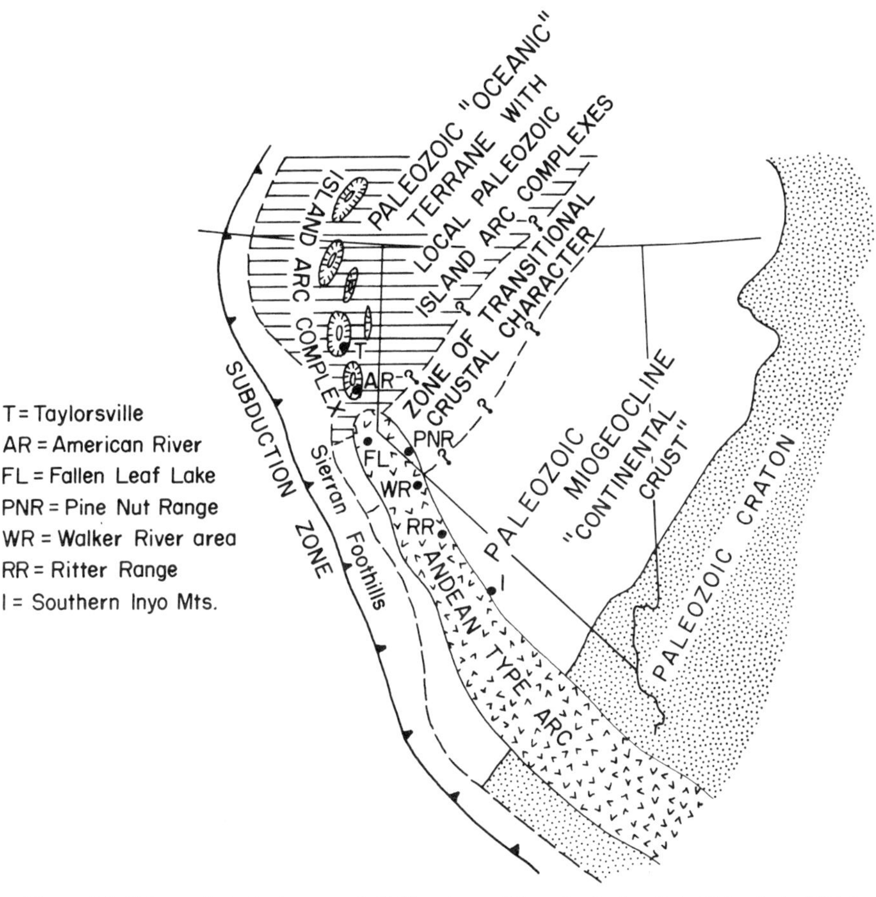

Figure 15. Schematic tectonic map of California and Nevada during Late Triassic to Middle Jurassic time (modified from Burchfiel and Davis, 1972, Fig. 7). Note that base map is not palinspastic for Cenozoic deformation.

delineated by sheared serpentinite and complex or chaotic deformation of older rocks. The youngest rocks cut by the fault zones are Oxfordian-Kimmeridgian (Late Jurassic), but the faults may have been active significantly earlier. Some of the fault zones are truncated by plutons that have yielded Late Jurassic radiometric ages (Evernden and Kistler, 1970). These older fault zones and the older plutonic and volcanic rocks east of the Sierra Nevada crest are suggested to manifest a period of subduction that occurred before the subduction of the Franciscan Complex. This suggestion implies that the zone of subduction jumped westward by Late Jurassic time and that the earlier subduction zone and associated rocks were accreted to the North American continent. A more complete discussion of these fault zones is presented elsewhere (Schweickert and Cowan, 1975).

Volcanic and plutonic rocks of the older, pre-Cretaceous magmatic arc may have foundered beneath voluminous outpourings of volcanic rocks during the construction of the younger, Late Jurassic to Late Cretaceous arc. If, as envisioned by Hamilton and Myers (1967), batholiths are emplaced beneath a thin cover of their own volcanic ejecta and frequently intrude just slightly older erupted rocks, the Cretaceous magmatic arc may have been constructed partly atop, partly adjacent to, and partly through the granitic and volcanic rocks of the pre-Cretaceous magmatic arc, with subsequent erosion accounting for the anomalous "deeper" structural level of the younger magmatic arc. Evidence of minor to moderate contact metamorphism of the older plutons, together with the resetting of ages of older plutons by younger intrusions (discussed above) probably reflects this superposition.

References Cited

Abbott, E. W., 1971, Stratigraphy and petrology of the Mesozoic volcanic rocks of southeastern California: Geol. Soc. America Abs. with Programs, v. 3, p. 69.

Bailey, E. H., Blake, M. C., Jr., and Jones, D. L., 1970, On-land Mesozoic oceanic crust in California Coast Ranges: U.S. Geol. Survey Prof. Paper 700C, p. C70-C81.

Bateman, P. C., 1961, Granitic formations in the east-central Sierra Nevada near Bishop, California: Geol. Soc. America Bull., v. 72, p. 1521-1538.

Bateman, P. C., and Dodge, F.C.W., 1970, Variations of major chemical constituents across the central Sierra Nevada batholith: Geol. Soc. America Bull., v. 81, p. 409-420.

Bateman, P. C., Clark, L. D., Huber, N. K., Moore, J. G., and Rinehart, C. D., 1963, The Sierra Nevada batholith—A synthesis of recent work across the central part: U.S. Geol. Survey Prof. Paper 414-D, p. D1-D46.

Best, M. G., 1963, Petrology of the Guadeloupe igneous complex, southwestern Sierra Nevada foothills, California: Jour. Petrology, v. 4, p. 223-259.

Burchfiel, B. C., and Davis, G. A., 1972, Structural framework and evolution of the southern part of the Cordilleran orogen, western United States: Am. Jour. Sci., v. 272, p. 97-118.

Castor, S. B., and McKee, E. H., 1975, Mesozoic intrusions, Pine Nut and Buckskin Ranges, western Nevada: Geol. Soc. America Abs. with Programs, v. 7, p. 303.

Cater, F., 1969, The Cloudy Pass epizonal batholith: Geol. Soc. America Spec. Paper 116, 52 p.

Chayes, F., 1969, The chemical composition of Cenozoic andesite *in* Andesite Conf., Eugene and Bend, Oreg., 1968, Proc. (Internat. Upper Mantle Proj. Sci. Rept. 16): Oregon Dept. Geol. and Mineral Industries Bull. 65, p. 1-11.

Clark, L. D., 1964, Stratigraphy and structure of parts of the western Sierra Nevada metamorphic belt: U.S. Geol. Survey Prof. Paper 410, 70 p.

Clark, L. D., Imlay, R. W., McMath, V. E., and Silberling, N. J., 1962, Angular unconformity between Mesozoic and Paleozoic rocks in the northern Sierra Nevada, California: U.S. Geol. Survey Prof. Paper 450B, p. B15-B19.

Clark, M. M., 1968, Pleistocene glaciation of the drainage of the West Walker River, Sierra Nevada, California [Ph.D. dissert.]: Stanford, Calif., Stanford Univ., 130 p.

Cowan, D. S., 1974, Deformation and metamorphism of the Franciscan subduction zone complex northwest of Pacheco Pass, California: Geol. Soc. America Bull., v. 85, p. 1623-1634.

Crickmay, C. H., 1933, Mount Jura investigation: Geol. Soc. America Bull., v. 44, p. 895-926.

Crowder, D. F., McKee, E. H., Ross, D. C., and Krauskopf, K. B., 1973, Granitic rocks of the White Mountains area, California-Nevada: Age and regional significance: Geol. Soc. America Bull., v. 84, p. 285-296.

Dickinson, W. R., 1970, Relations of andesites, granites, and derivative sandstones to arc-trench tectonics: Rev. Geophysics and Space Physics, v. 8, p. 813-860.

Dickinson, W. R., and Rich, E. I., 1972, Petrologic intervals and petrofacies in the Great Valley sequence, Sacramento Valley: Geol. Soc. America Bull., v. 83, p. 3007-3024.

Diller, J. S., 1908, Geology of the Taylorsville region, California: U.S. Geol. Survey Bull. 353, 128 p.

Dunne, G. C., 1971, Petrology of the Pat Keyes pluton, Inyo Mountains, California, and its relation to the Sierra Nevada batholith: Geol. Soc. America Abs. with Programs, v. 3, p. 113.

Ernst, W. G., 1970, Tectonic contact between the Franciscan mélange and the Great Valley sequence—Crustal expression of a late Mesozoic Benioff zone: Jour. Geophys. Research, v. 75, p. 886-902.

Evernden, J. F., and Kistler, R. W., 1970, Chronology of emplacement of Mesozoic batholithic complexes in California and western Nevada: U.S. Geol. Survey Prof. Paper 623, 42 p.

Fiske, R. W., Hopson, C. A., and Waters, A. C., 1963, Geology of Mount Rainier National Park: U.S. Geol. Survey Prof. Paper 444, 93 p.

Forbes, R. B., Ray, D. K., Katsura, T., Matsumoto, H., Haramura, H., and Furst, M. J., 1969, The comparative chemical composition of continental vs. island arc andesites in Alaska, *in* Andesite Conf., Eugene and Bend, Oreg., 1968, Proc. (Internat. Upper Mantle Prog. Sci. Rept. 16): Oregon Dept. Geology and Mineral Industries Bull. 65, p. 111-120.

Hall, W. E., and MacKevett, E. M., Jr., 1962, Geology and ore deposits of the Darwin quadrangle, Inyo County, California: U.S. Geol. Survey Prof. Paper 368, 83 p.

Halsey, J. H., 1953, Geology of parts of the Bridgeport, California, and Wellington, Nevada, quadrangles [Ph.D. dissert.]: Berkeley, Univ. California.

Hamilton, W., 1969a, The volcanic central Andes—A modern model for the Cretaceous batholiths of western North America, *in* Andesite Conf., Eugene and Bend, Oreg., 1968, Proc. (Internat. Upper Mantle Proj. Sci. Rept. 16): Oregon Dept. Geology and Mineral Industries Bull. 65, p. 175-184.

——1969b, Mesozoic California and the underflow of the Pacific mantle: Geol. Soc. America Bull., v. 80, p. 2409-2430.

Hamilton, W., and Myers, W. B., 1967, The nature of batholiths: U.S. Geol. Survey Prof. Paper 554C, p. C1-C30.

Harker, A., 1904, The Tertiary igneous rocks of Skye: Great Britain Geol. Survey Memoirs.

Jackson, E. D., 1961, Primary textures and mineral associations in the ultramafic zone of the Stillwater complex, Montana: U.S. Geol. Survey Prof. Paper 358, 106 p.

Kistler, R. W., 1966a, Structure and metamorphism in the Mono Craters quadrangle, Sierra Nevada, California: U.S. Geol. Survey Bull. 1221E, 53 p.

——1966b, Geologic map of the Mono Craters quadrangle, Mono and Tuolumne Counties, California: U.S. Geol. Survey Geol. Quad. Map GQ-462, scale 1:62,500.

——1974, Phanerozoic batholiths in western North America: Annu. Rev. Earth and Planetary Sci., v.2, p. 403-418.

Kistler, R. W., Evernden, J. L., and Shaw, H. R., 1971, Sierra Nevada plutonic cycle: Pt. I. Origin of composite granitic batholiths: Geol. Soc. America Bull., v. 82, p. 853-868.

Kistler, R. W., and Peterman, Z. E., 1973, Variations in Sr, Rb, K, Na, and initial Sr^{87}/Sr^{86} in Mesozoic granitic rocks and intruded wall rocks in central California: Geol. Soc. America Bull., v. 84, p. 3489-3512.

Knopf, A., 1918, A geologic reconnaissance of the Inyo Range and the east slope of the southern Sierra Nevada, California: U.S. Geol. Survey Prof. Paper 110, 130 p.

Koenig, J. B., 1963, Walker Lake Sheet, geologic map of California: California Div. Mines and Geology.

Lanphere, M. A., 1971, Age of the Mesozoic oceanic crust in the California Coast Ranges: Geol. Soc. America Bull., v. 82, p. 3209-3212.

Lanphere, M. A., and Reed, B. L., 1973, Timing of Mesozoic and Cenozoic plutonic events in circum-Pacific North America: Geol. Soc. America Bull., v. 84, p. 3773-3782.

Loomis, A. A., 1961, Petrology of the Fallen Leaf Lake area, California [Ph.D. dissert.]: Stanford, Calif., Stanford Univ., 121 p.

——1966, Contact metamorphic reactions and processes in the Mt. Tallac roof remnant, Sierra Nevada, California: Jour. Petrology, v. 7, p. 221-245.

McAllister, J. F., 1956, Geology of the Ubehebe Peak quadrangle, California: U.S. Geol. Survey Geol. Quad. Map GQ-95, scale 1:62,500.

McKee, E. H., and Nash, D. B., 1967, Potassium-argon ages of granitic rocks in the Inyo batholith, east-central California: Geol. Soc. America Bull., v. 78, p. 669-680.

McMath, V. E., 1966, Geology of the Taylorsville area, northern Sierra Nevada: Calif. Div. Mines and Geology Bull. 190, p. 173-183.

Merriam, C. W., 1963, Geology of the Cerro Gordo mining district, Inyo County, California: U.S. Geol. Survey Prof. Paper 408, 83 p.

Moore, J. G., 1963, Geology of the Mount Pinchot quadrangle, southern Sierra Nevada, California: U.S. Geol. Survey Bull. 1130, 150 p.

Noble, D. C., 1962, Mesozoic geology of the southern Pine Nut Range, Douglas County, Nevada [Ph.D. dissert.]: Stanford, Calif., Stanford Univ.

Nockolds, S. R., 1954, Average chemical compositions of some igneous rocks: Geol. Soc. America Bull., v. 65, p. 1007-1032.

Ojakangas, R. W., 1968, Cretaceous sedimentation, Sacramento Valley, California: Geol. Soc. America Bull., v. 79, p. 973-1008.

Rinehart, C. D., and Ross, D. C., 1964, Geology and mineral deposits of the Mount Morrison quadrangle, Sierra Nevada, California: U.S. Geol. Survey Prof. Paper 385, 106 p.

Rinehart, C. D., Ross, D. C., and Huber, N. K., 1959, Paleozoic and Mesozoic fossils in a thick stratigraphic section in the eastern Sierra Nevada: Geol. Soc. America Bull., v. 70, p. 941-945.

Rogers, J.J.W., Burchfiel, B. C., Abbott, E. W., Anepohl, J. K., Ewing, A. H., Koehnken, P. J., Novitsky-Evans, J. M., and Talukdar, S. C., 1974, Paleozoic and lower Mesozoic volcanism and continental growth in the western United States: Geol. Soc. America Bull., v. 85, p. 1913-1924.

Ross, D. C., 1965, Geology of the Independence quadrangle, Inyo County, California: U.S. Geol. Survey Bull. 1181-O, 64 p.

——1969, Descriptive petrography of three large granitic bodies in the Inyo Mountains, California: U.S. Geol. Survey Prof. Paper 601, 47 p.

Schweickert, R. A., and Cowan, D. S., 1975, Early Mesozoic tectonic evolution of the western Sierra Nevada, California: Geol. Soc. America Bull., v. 86, p. 1329-1336.

Sherlock, D. G., and Hamilton, W., 1958, Geology of the north half of the Mount Abbott quadrangle, Sierra Nevada, California: Geol. Soc. America Bull., v. 69, p. 1245-1268.

Silberling, N. J., 1973, Geologic events during Permian-Triassic time along the Pacific margin of the United States, *in* Logan, A., and Hills, L. V., eds., The Permian and Triassic Systems and their mutual boundary: Calgary, Alberta Soc. Petroleum Geologists, p. 345-362.

Silberman, M. C., and McKee, E. H., 1971, Periods of plutonism in north-central Nevada, a section on gold-bearing deposits in north-central Nevada and southwestern Idaho: Econ. Geology, v. 66, p. 17-20.

Smith, J. G., McKee, E. H., Tatlock, D. B., and Marvin, R. F., 1971, Mesozoic granitic rocks in northwestern Nevada: A link between the Sierra Nevada and Idaho batholiths: Geol. Soc. America Bull., v. 82, p. 2933-2944.

Speed, R. C., 1963, Layered picrite-anorthositic gabbro sheet, West Humboldt Range, Nevada: Mineralog. Soc. America Spec. Paper 1, p. 69-77.

——1973, Excelsior and Pablo Formations, western Nevada: Problems and progress in analysis:

Geol. Soc. America Abs. with Programs, v. 5, p. 109-110.

Stanley, K. O., Jordan, W. M., and Dott, R. H., Jr., 1971, New hypothesis of Early Jurassic paleogeography and sediment dispersal for the western United States: Am. Assoc. Petroleum Geologists Bull., v. 55, p. 10-19.

Stuart-Alexander, D. E., 1966, Contrasting deformation of Paleozoic and Mesozoic rocks near Sierra City, northern Sierra Nevada, California [Ph.D. dissert.]: Stanford, Calif., Stanford Univ.

Suppe, J., 1969, Times of metamorphism in the Franciscan terrain of the northern Coast Ranges, California: Geol. Soc. America Bull., v. 80, p. 135-142.

——1973, Geology of the Leech Lake Mountain-Ball Mountain region, California: California Univ. Pubs. Geol. Sci. Bull. 107, 82 p.

Sylvester, A. G., Nelson, C. A., and Barrows, A., 1969, Monzonite in the Inyo batholith, east-central California: Geol. Soc. America Abs. with Programs, v. 1, p. 67.

Vance, J. A., 1969, On synneusis: Contr. Mineralogy and Petrology, v. 24, p. 7-29.

Wager, L. R., and Brown, G. M., 1967, Layered igneous rocks, San Francisco, W. H. Freeman and Company, 572 p.

Waters, A. C., 1962, Basalt magma types and their tectonic associations: Pacific northwest of the United States: Am. Geophys. Union Mon. 6, p. 158-170.

Manuscript Received by the Society May 2, 1974
Revised Manuscript Received November 3, 1975
Manuscript Accepted January 12, 1976